Molecular Thermodynamics and Transport Phenomena

Complexities of Scales in Space and Time

McGraw-Hill Nanoscience and Technology Series

KENNETH GILLEO • *MEMS/MOEM Packaging: Concepts, Design, Materials, and Processes*

DARREN L. HITT • *Microfluidics: Experimental Techniques and Applications*

JOSEPH H. KOO • *Polymer Nanocomposites: Processing, Characterization and Applications*

NICOLAE O. LOBONTIU • *Mechanical Design of Microresonators: Modeling and Applications*

OMAR MANASREH • *Semiconductor Heterojunctions and Nanostructures*

ROBERTO PAIELLA • *Intersubband Transitions In Quantum Structures*

MICHAEL H. PETERS • *Molecular Thermodynamics and Transport Phenomena*

JENS W. TOMM AND JUAN JIMENEZ • *Quantum-Well High-Power Laser Arrays*

Molecular Thermodynamics and Transport Phenomena

Complexities of Scales in Space and Time

Michael H. Peters, Ph.D.
Virginia Commonwealth University

McGraw-Hill

New York Chicago San Francisco Lisbon London
Madrid Mexico City Milan New Delhi San Juan
Seoul Singapore Sydney Toronto

The McGraw·Hill Companies

Library of Congress Cataloging-in-Publication Data

Peters, Michael H.
Molecular thermodynamics and transport phenomena : complexities of scales in space and time / Michael H. Peters.
p. cm.
Includes index.
ISBN 0-07-144561-7
1. Thermodynamic equilibrium—Textbooks. 2. Transport theory—Textbooks. 3. Space and time. 4. Mechanics. I. Title.
QC318.T47P47 2005
530.13′8—dc22

2004065564

1 2 3 4 5 6 7 8 9 0 DOC/DOC 0 1 0 9 8 7 6 5

ISBN 0-07-144561-7

The sponsoring editor for this book was Kenneth P. McCombs and the production supervisor was Sherri Souffrance. It was set in Century Schoolbook by International Typesetting and Composition. The art director for the cover was Margaret Webster-Shapiro.

Printed and bound by RR Donnelley.

This book is printed on recycled, acid-free paper containing a minimum of 50% recycled, de-inked fiber.

In memory of Debbie and Harold

Contents

Preface

This book is intended for upper-level undergraduates and first-year graduate students in science and engineering. It is basically an introduction to the molecular foundations of thermodynamics and transport phenomena presented in a unified manner. The mathematical and physical science level sophistication are at the upper undergraduate level, and students who have been exposed to vector calculus, differential equations, and calculus-based physics should be adequately prepared to handle the material presented. There is also sufficient advanced material in each chapter or topic for first-year graduate students in science or engineering.

The basic foundations of both (equilibrium) thermodynamics and transport phenomena lie in the descriptions and treatment of large collections of interacting molecules and atoms. From this viewpoint, there are only minor differences in fundamentally describing the behavior of systems at equilibrium versus nonequilibrium, and these differences are immersed in the mathematical identification of length and time scales appropriate for any particular system. The importance of understanding the complexities of length and time scales is ubiquitous throughout science and engineering and is not unique to thermodynamics and transport phenomena. As such, this book will provide systematic and methodological examples of the analysis of length and time scales and its implications for physical behavior.

The book begins by providing the necessary background for the mechanics of interacting particles in Chapter 1. Atomic and molecular interaction forces are also presented in this introductory chapter. Chapter 2 develops the formal and general statistical representation of a collection of interacting particles through the derivation of the Liouville equation, the fundamental equation of classical statistical mechanics. Chapter 3 examines reduced forms of the Liouville equation, the simplest of which is the famous Boltzmann transport equation. In Chapter 4, the Liouville equation is used as a starting point to develop the equilibrium properties of matter. The necessary characteristics of global equilibrium behavior, such as time independence and spatial homogeneity, are directly used to solve the Liouville equation under these conditions. The basic thermodynamic functions are also defined in this chapter and expressions are formally obtained for both gas and liquid systems. Example calculations of thermodynamic functions for specific chemical systems are also given in the chapter. In Chapter 5, we turn to nonequilibrium behavior and derive the most general forms of the so-called transport equations directly from the Liouville equation. We explicitly show how the transport equations simply represent the nonequilibrium counterparts of the thermodynamic functions introduced in Chapter 4, thus establishing a direct and comprehensive link between equilibrium thermodynamics and transport phenomena; this link includes both equilibrium and nonequilibrium entropy. Finally, in Chapter 6 we examine some of the complexities of length and time scales by analyzing the mathematical solutions of reduced forms of the Liouville equation for describing the nonequilibrium behavior of both gas and liquid systems.

This book could be used in a variety of ways: (1) as a one-semester advanced undergraduate or graduate course in equilibrium thermodynamics by covering material from Chapters 1 through 4; (2) as a one-semester advanced undergraduate or graduate course in transport phenomena by covering Chapters 1 through 3, 5, and 6; or (3) as a two-semester course sequence in thermodynamics and transport phenomena.

I want to gratefully thank all my instructors in science, engineering, and mathematics who contributed in some way to the compilation of the knowledge contained in this book. Hopefully, I have done them justice! I want to specifically thank Dr. Raj Rajagopalan, Dr. Cynthia S. Hirtzel, and Dr. Howard Brenner who stimulated my original interest in statistical mechanics. Last but not least, special thanks to my mother Louise, sister Cheryl, and my very understanding spouse Karen and daughter Emily for their constant support.

Michael H. Peters

Notes on Notation

All variables are defined where they are introduced in the text.

1. Vectors and tensors are denoted in boldface, e.g., $\mathbf{v}$ is a velocity vector, $\mathbf{P}$ is a pressure tensor, etc.
2. Unit vectors are denoted as $\mathbf{e}_i$, where i denotes the direction.
3. Specific (per mole) thermodynamic properties are denoted by a hat over the symbol, e.g., $\hat{V}$ denotes the specific volume or the volume per mole.
4. Multidimensional integrals are denoted using boldface differential symbols under a single integral sign, e.g., $\int d\mathbf{r}$ denotes the actual three dimensional integral $\int\int\int dx\, dy\, dz$

Likewise, we use a single integral sign $\int d\mathbf{r}^N$ to denote the actual $3N$ multidimensional integral $\int \cdots \int d\mathbf{r}_1 \cdots d\mathbf{r}_N$

Chapter

1

Introduction to Statistical Mechanics and the Classical Mechanics of Interacting Particles

"There are so many things that you really should know
And that's why I'm bothering telling you so.
You should know the first names of the Fuddnuddler Brothers
Who like to pile each on the heads of others."

–THEODOR SEUSS GEISEL,
Oh Say Can You Say?

1.1 Introduction to Classical Statistical Mechanics and a Unification of Equilibrium and Nonequilibrium Analysis

Broadly speaking, all physical systems may be globally classified as being either in a state of (global) equilibrium or (global) nonequilibrium. Here we parenthetically use the word "global" to mean the entire system to avoid any confusion with the word "local" that implies a small part or point within the entire system. A system in *(global) equilibrium* is one that exhibits no variations in space or time. A container enclosing a fixed number of molecules in a gas, liquid, or solid state completely isolated from any external forces or interactions with its surroundings will maintain this state for an indefinite period. Such a system, which is completely isolated from its surroundings, is said to be in a state of (global) equilibrium, as it exhibits no variations with respect to time or space.

Systems are said to be in *(global) nonequilibrium* states due to an observed variation in their properties, such as temperature, pressure, and concentration, with respect to time or space. For example, the temperature along a solid bar may vary from one end to the other, which classifies it as being in a (global) nonequilibrium state. Similarly, a system temperature may be uniform in space but vary with respect to time. In general, (global) nonequilibrium systems that do not change with time and only vary in space are further classified as being in a so-called *steady-state* condition. Any nonequilibrium system that changes with time is said to be in an *unsteady-state condition*. Henceforth, we will drop the notation "(global)" as is customarily done, and it will be implied that when we speak of equilibrium or nonequilibrium we mean in a global sense unless specifically noted otherwise.

Now, the macroscopic behavior† of all systems, whether in equilibrium or nonequilibrium states, is classically‡ described in terms of the thermodynamic variables: pressure $\mathbf{P}$, temperature T, specific volume $\hat{V}$ (volume per mole), specific internal energy $\hat{U}$ (energy per mole), specific entropy $\hat{S}$ (entropy per mole), concentration or chemical potential μ, and velocity $\mathbf{v}$.§ In nonequilibrium states, these variables change with respect to space and/or time, and the subject matter is called *nonequilibrium thermodynamics*. When these variables do not change with respect to space or time, their prediction falls under the subject matter of *equilibrium thermodynamics*. As a matter of notation, we would indicate a nonequilibrium variable such as entropy by $\hat{S}\,(\mathbf{r}, t)$, where $\mathbf{r}$ is a vector that locates a particular region in space (locator vector) and t is the time, whereas the equilibrium notation would simply be $\hat{S}$.

Today, the subject matter of nonequilibirum thermodynamics is usually encompassed in a variety of courses on so-called heat transfer, fluid mechanics, transport phenomena, nonlinear hydrodynamics, and the like. These courses often seem disconnected from one another and certainly quite disconnected from courses in so-called thermodynamics, which usually means equilibrium thermodynamics. However, what we will show in this text is that all these particular subjects have a common fundamental basis of description, and they actually encompass

---

†By macroscopic we mean some type of local average over a large group of molecules of the system.

‡By "classically" we mean according to the laws of classical physics, as opposed to the laws of quantum physics. This text will not consider the quantum physical aspects of equilibrium or nonequilibrium thermodynamics.

§By definition, the velocity is always zero for a system in equilibrium, so it is not included as variable in conventional (equilibrium) thermodynamic texts.

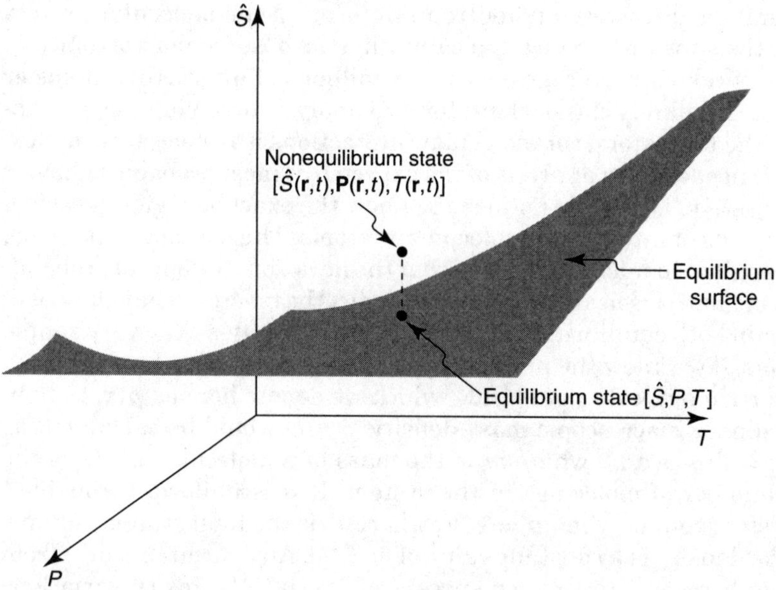

Figure 1.1 Phase diagram of a substance depicting the differences between equilibrium and nonequilibrium states.

a general set of macroscopic variables that can be developed simultaneously from one common basis. This integration and generalization requires that we think about all physical systems differently, i.e., beginning with a molecular description of matter.

In addition to a common set of variables, equilibrium and nonequilibrium systems are related in a very close dynamical manner. As shown in Fig. 1.1, the set of all equilibrium states available to a system can be represented by an equilibrium surface, such as specific entropy $\hat{S}$, temperature T, and specific volume $\hat{V}$. Most nonequilibrium states of practical interest represent perturbations from a particular global or local equilibrium state, as depicted in Fig. 1.1. For most systems, the perturbation from an equilibrium state is caused by some external means, such as a force or disturbance acting on the system from the surroundings; and if this external force or disturbance is removed, the system has a tendency to resort back to its equilibrium state. In addition to this preferred direction of change for nonequilibrium states, known as irreversibility, and because of the close dynamical relation between equilibrium and nonequilibrium states, a unified analysis and treatment is not only possible but highly desirable in terms of developing an intrinsic understanding of physical phenomena in general.

The role of statistical mechanics is in the prediction of the behavior of the so-called macroscopic variables mentioned above, such as temperature, pressure, and entropy, in terms of the molecular properties of the substance. Since typically what we observe macroscopically is some kind of an average over many millions of interacting atoms or molecules, a purely classical mechanical analysis involving the description of the total atomic or molecular interactions is hopelessly complex. What is needed is an analysis of the expected or most probable behavior of the system, i.e., we do not need to know the exact behavior of each of the millions of interacting molecules or atoms. Thus, broadly speaking, statistical mechanics merges together the laws and concepts of probability theory and classical mechanics to predict the macroscopic behavior of matter in both equilibrium and nonequilibrium states. As a very simple example, if we knew the probability of finding any molecule of a system at a particular location and time, which we denote here as $p(\mathbf{r}, t)$, then the ordinary macroscopic mass density, $\rho(\mathbf{r}, t)$ would be calculated as $\rho(\mathbf{r}, t) = mNp(\mathbf{r}, t)$, where m is the mass of a molecule and N is the total number of molecules in the system. If p is uniform throughout the system volume, then $p = 1/V$, where V is the total system volume, and the density is a constant value of mN/V. Any variations in p from $1/V$ with respect to time or space are a result of density variations from this uniform value. Predicting these density variations would be critically important in describing the overall behavior of the system, including changes of phase from, say, a gas to liquid state or changes in other physical properties such as viscosity and thermal conductivity. How exactly to determine the probability function $p(\mathbf{r}, t)$ requires the use of probability theory and classical mechanics together. This fundamental starting point will be introduced in Chap. 2, and for now we need to learn more about the molecular level description of matter and review as much classical mechanics as is necessary for further development of our unified approach to the equilibrium and nonequilibrium description of matter.

1.2 Molecular Interactions in Gases, Liquids, and Solids and the Nature of Intermolecular (Interatomic) Interaction Forces

The nature and degree of molecular interactions in substances dictates their physical existence as a gas, liquid, solid, or combinations thereof. In general, the interaction force between any two molecules of a system decays rapidly with increasing separation distance. In gases, the intermolecular spacing is large (much greater than the molecular diameter) and, consequently, intermolecular interactions are weak. In liquids and solids, on the other hand, the intermolecular spacing is small (on the order of the molecular diameter) and, consequently, the intermolecular

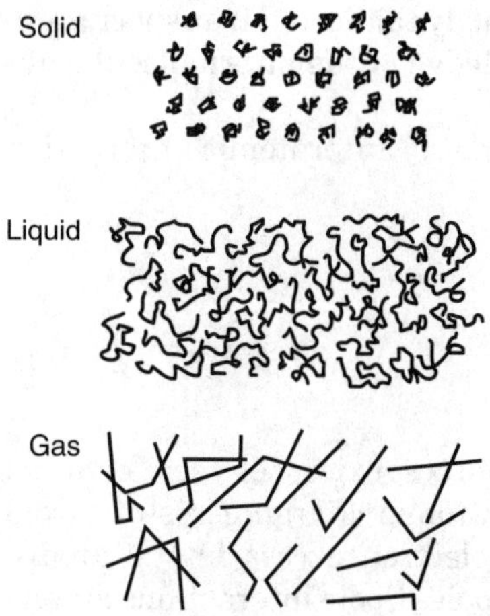

Figure 1.2 Trajectories of interacting molecules in gases, liquids, and solids.

interaction forces are relatively strong. The trajectories of molecules in these different phases are illustrated in Fig. 1.2.

As shown in Fig. 1.2, the molecules of any substance undergo seemingly "random" motion. This randomness is a manifestation of the multitude of interactions between the molecules. Actually, each molecule moves deterministically; for example, according to Newton's second law of motion, $\mathbf{F}_i = d(m_i\mathbf{v}_i)/dt$, where $\mathbf{F}_i$ is the total force acting on molecule (i) due to any external forces, such as gravity, present and the intermolecular interaction forces due to the surrounding molecules.

As we will see in much more detail later in Chap. 4, the temperature of any substance is a measure of the molecular randomness or, more specifically, the variance in the molecular velocity distribution function. The greater the variability or spread in molecular velocities, the greater the temperature, and vice versa. Heating and cooling of a substance represents a *transfer* of energy to or from the substance, respectively. As will be seen in greater detail in Chap. 5, the energy transferred in this case is very specifically defined in terms of molecular parameters involving both the average molecular kinetic energies, which are related to temperature, and the intermolecular interaction potential energies, which are introduced below. Similarly, as will be shown in detail in Chap. 5, the pressure of any substance is rigorously defined in terms of the net molecular momentum flux through an imaginary plane in the substance. This transfer of momentum can be interpreted as a force that can act on any external body or surface placed in the transfer pathway, such as the walls containing a substance. In general, all thermodynamic (equilibrium) and transport (nonequilibrium) properties are rigorously defined in terms of the molecular properties of the substance. Our goals in this text will be to develop these relationships and to show how the thermodynamic and transport properties can be determined from the knowledge of the intermolecular interaction forces.

In general, elementary intermolecular or interatomic interaction forces can be classified as follows:

1. Electrostatic forces
2. Van der Waals forces
3. Born repulsive forces
4. Covalent bonding

Coulombic electrostatic interaction forces exist between charged molecules or atoms. The molecular or atomic charging results from the condition of electron deficiency or electron excess in the atoms. Other common electrostatic forces are dipole-dipole interactions among polar molecules, such as water, and higher-order multipole interactions. Van der Waals forces are associated with oscillating, time dependent electric fields created by the "wave" or probabilistic behavior of electrons in atoms. The van der Waals forces between atoms are attractive in nature and vary as r^{-7}, where r is the interatomic separation distance. Born repulsive forces result when two atoms or molecules approach extremely close to one another at distances slightly larger than σ, where σ is the atomic (molecular) diameter. At these very small separation distances, the electron shells of two interacting atoms or molecules can partly penetrate one another. The nuclear charge interatomic interactions, however, which are no longer fully screened by the electrons, cause strong repulsive forces between the atoms or molecules, preventing further overlap. Repulsive forces are short-range forces compared to van der Waals forces. Covalent bonding results when an electron is shared by two adjoining atoms. Covalent bonding is important when chemical reactions take place (bond breaking and forming) and will not be considered explicitly in this text.

Consider the force (type 1, 2, or 3) acting on atom or molecule (1) by atom or molecule (2) as shown in Fig. 1.3. For *central forces*, we have

$$\mathbf{F} = F_r \mathbf{e}_r \tag{1.1}$$

where $\mathbf{e}_r = \mathbf{r}/|\mathbf{r}| = \mathbf{r}/r$ is a unit vector pointing from atom or molecule (2) to atom or molecule (1) (Fig. 1.3). If $F_r < 0$, the force acting on (1) is attractive, and if $F_r > 0$, the force is repulsive as shown in Fig. 1.3. The potential of the force is defined according to

$$F_r(r) = -\frac{du(r)}{dr} \tag{1.2}$$

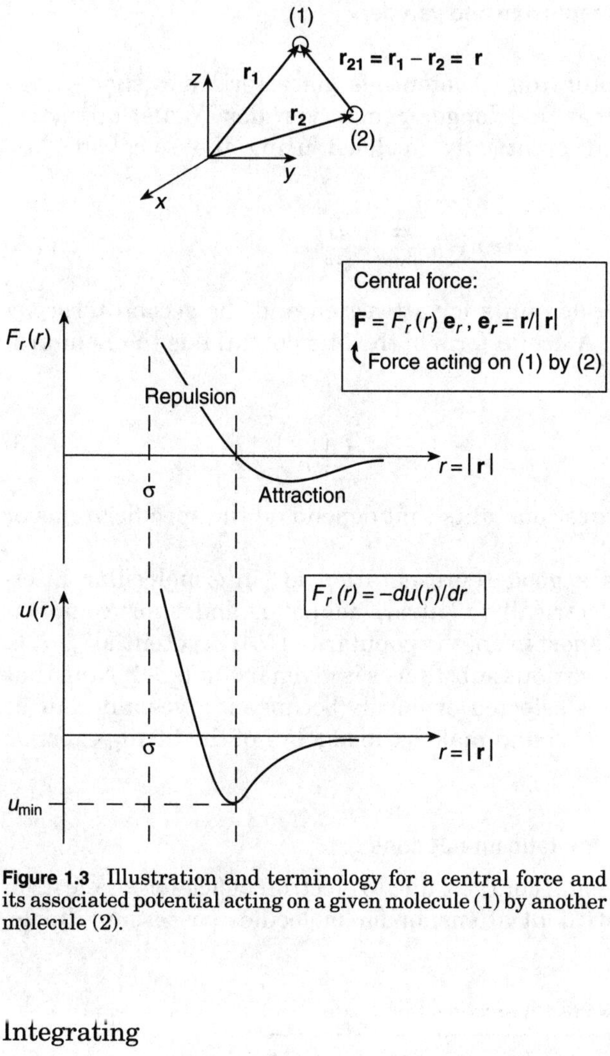

Figure 1.3 Illustration and terminology for a central force and its associated potential acting on a given molecule (1) by another molecule (2).

Integrating

$$-\int_{\infty}^{r} F_r(r')\,dr' = \int_{\infty}^{r} du = u(r) - u(\infty) = u(r) \tag{1.3}$$

where we have arbitrarily set $u(\infty) = 0$ as the reference potential. Thus

$$u(r) = \int_{r}^{\infty} F_r(r')\,dr' \tag{1.4}$$

This is sketched in Fig. 1.2.

1.2.1 Modeling Born repulsive and van der Waals forces

Intermolecular (or interatomic) potentials characterized by short-range (Born) repulsive forces and longer range (van der Waals) attractive forces are often mathematically modeled using the so-called "Mie potential" expression

$$u(r) = -\frac{A}{r^m} + \frac{B}{r^n} \tag{1.5}$$

where the first term accounts for attraction and the second term accounts for repulsion. A useful form of the Mie potential is the Lennard-Jones (L-J) potential

$$u(r) = 4\epsilon\left[-\left(\frac{\sigma}{r}\right)^p + \left(\frac{\sigma}{r}\right)^q\right] \tag{1.6}$$

where ϵ, σ, p and q are constants that depend on the specific atoms or molecules of interest.

The L-J model is a good representation for intermolecular interactions involving electrically neutral, nonpolar, and short/compact-shaped molecules. A short list of the popular L-J (6-12) potential ($p = 6$, $q = 12$) constants for various substances is given in Table 1.1. Note that the value of $q = 2p$ is selected primarily because it gives a decent fit of the experimental data and makes the algebra of the fitting exercise simpler.

1.2.2 Models for electrostatic interactions

For molecules possessing charges, due to electron deficiency or electron excess in their constituent atoms, or for molecules possessing dipole,

TABLE 1.1 Lennard-Jones (6-12) Potential Constants

Substance	ϵ/K	σ(Å)
N_2	95.05	3.698
O_2	118	3.46
CH_4	148.2	3.817
CO_2	189	4.486
CCl_4	327	5.881
C_2H_4	185	4.221
C_6H_6	440	5.270

SOURCE: Appendix, Table I-A, J.O. Hirschfelder, C.F. Curtiss, and R.B. Bird, *Molecular Theory of Gases and Liquids*, Wiley, New York, 1960. Table I-A contains an extensive listing. Note that k is Boltzmann's constant, $k = 1.380 \times 10^{-23}$ Joule · molecule^{-1}K^{-1}, and $1\text{Å} \equiv 10^{-10}$m.

quadrapole, or higher moments, due to shifting of electron positioning (in a time-averaged sense) between bonded atoms, the L-J potential model alone is insufficient to describe intermolecular interactions.

The simplest electrostatic interaction is the Coulomb potential between two point charges q_1 and q_2, located a distance r apart:

$$u(r) = \frac{q_1 q_2}{r} \tag{1.7}$$

where q_1 and q_2 are given in electrostatic units (1 e.s.u $\equiv$ 1 dyn$^{1/2}\cdot$cm). The Coulomb potential, along with a Born repulsive potential, is often a good model for simple ionic interactions such as interactions between Na^+ and Cl^-.

For molecules possessing dipole, or higher moments, such as water, and for molecules with localized charge groups, such as proteins and fatty acids, the structural form of the molecule must be taken into account. A complete presentation is beyond the scope of this writing and we refer the interested reader to the Further Reading Section at the end of this chapter. However, in the example below, the dipole-dipole interaction potential is derived, which has useful applications. The introduction of molecular structure requires the use of the so-called Euler angles that give the orientation of the molecule relative to a fixed or laboratory reference frame. The Euler angles are described below, which for readers unfamiliar to them should be consulted before studying the dipole-dipole interaction example that follows.

Interlude 1.1 Mathematical Essentials: The Euler Angles A description of the orientation of an object relative to a space-fixed, laboratory reference frame is not an easy task. Consider the simple example of describing the orientation of a rectangular block as shown in Fig. 1.4. We could define rotations of the block about a space-fixed cartesian coordinate system as ϕ_x, ϕ_y, and ϕ_z, to obtain a particular orientation. Unfortunately, it is easy to show that if we rotate the block through specified values of ϕ_x, ϕ_y, and ϕ_z, the final orientation of the block depends on the order in which we perform these rotations.†

Fortunately for us, the problem of assigning rotational angles in unambiguous ways has been mathematically resolved. There are several different sets of rotational angle representations, and we will consider only one of these here—the so-called Euler angles. Other representations can be found in the further list Reading Section at the end of this chapter.

†Try this with a book as an example. Begin with the book lying flat on a table and define a space-fixed cartesian coordinate system. Rotate the book around each axis by 90° in the order (ϕ_x, ϕ_y, and ϕ_z); then repeat the example by rotating in the order (ϕ_x, ϕ_z, and ϕ_y). Note how dramatically different the final orientations are in the two cases!

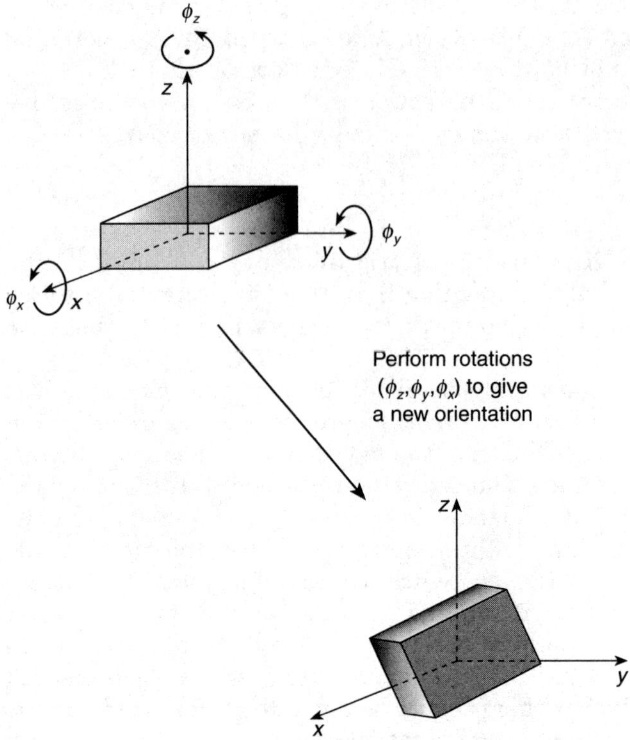

Figure 1.4 Rotation of a block around cartesian coordinate axes.

In order to succinctly see how the Euler angles are defined and used, first consider a simpler *two-dimensional* system shown in Fig. 1.5. The reference frame $(\bar{x}, \bar{y})$ is obtained by rotating the (x, y) frame counterclockwise through an angle θ. We wish to describe a fixed vector $\mathbf{A}$ in both frames. We have

$$\begin{aligned} \mathbf{A} &= A_x \mathbf{e}_x + A_y \mathbf{e}_y, \qquad (x, y) \text{ frame} \\ &= A_{\bar{x}} \mathbf{e}_{\bar{x}} + A_{\bar{y}} \mathbf{e}_{\bar{y}}, \qquad (\bar{x}, \bar{y}) \text{ frame} \end{aligned} \tag{1.8}$$

where A_x and A_y are the components of $\mathbf{A}$ in the (x, y) frame, and $A_{\bar{x}}$ and $A_{\bar{y}}$ are the components of $\mathbf{A}$ in the $(\bar{x}, \bar{y})$ frame. Now, we can easily relate these two sets of components as follows:

$$\begin{aligned} A_x &= \mathbf{e}_x \cdot \mathbf{A} = A_{\bar{x}}(\mathbf{e}_x \cdot \mathbf{e}_{\bar{x}}) + A_{\bar{y}}(\mathbf{e}_x \cdot \mathbf{e}_{\bar{y}}) \\ &= A_{\bar{x}} \cos\theta - A_{\bar{y}} \sin\theta \end{aligned} \tag{1.9}$$

and

$$\begin{aligned} A_y &= \mathbf{e}_y \cdot \mathbf{A} = A_{\bar{x}}(\mathbf{e}_y \cdot \mathbf{e}_{\bar{x}}) + A_{\bar{y}}(\mathbf{e}_y \cdot \mathbf{e}_{\bar{y}}) \\ &= A_{\bar{x}} \sin\theta + A_{\bar{y}} \cos\theta \end{aligned} \tag{1.10}$$

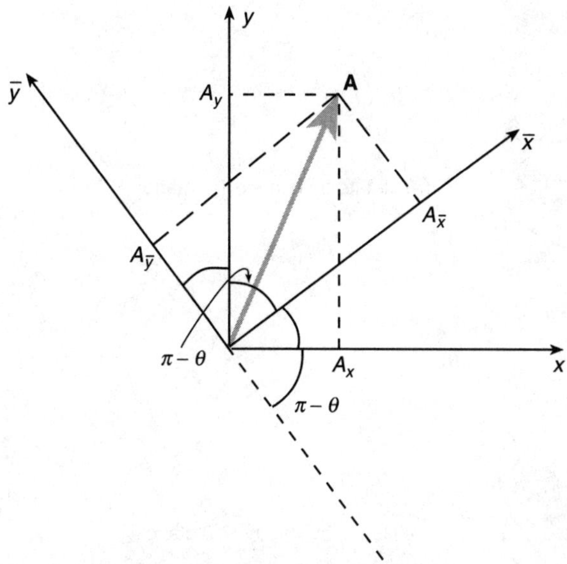

Figure 1.5 Components of a vector **A** in two reference frames rotated by an angle θ from each other.

In matrix notation

$$\begin{bmatrix} A_x \\ A_y \end{bmatrix} = \begin{bmatrix} \cos\theta & -\sin\theta \\ \sin\theta & \cos\theta \end{bmatrix} \begin{bmatrix} A_{\bar{x}} \\ A_{\bar{y}} \end{bmatrix} \tag{1.11}$$

or

$$A_i = \sum_{\bar{j}} e_{i\bar{j}} A_{\bar{j}} \tag{1.12}$$

where $i = x, y$ and the sum runs over all $\bar{j}$, $\bar{j} = \bar{x}, \bar{y}$, and $e_{i\bar{j}}$ are the matrix elements of the transformation matrix

$$\left[e_{i\bar{j}}\right] = \begin{bmatrix} \cos\theta & -\sin\theta \\ \sin\theta & \cos\theta \end{bmatrix} \tag{1.13}$$

Now, consider a *three-dimensional* system. We wish to describe a fixed vector **A** in the (x, y, z) system, which will be thrice rotated to a $(\bar{x}, \bar{y}, \bar{z})$ system. For ease of notation, we will label the (x, y, z) axes as (1, 2, 3) and the $(\bar{x}, \bar{y}, \bar{z})$ axes as $(\bar{1}, \bar{2}, \bar{3})$. All sets of axes will be assumed to represent orthogonal coordinate systems.

The first rotation of axes will be to rotate the (1, 2, 3) axes by an angle ϕ in the 1-2 plane as shown in Fig. 1.6*a*. We label these new axes as $(\hat{1}, \hat{2}, \hat{3})$ and the transformation matrix, by analogy with the two-dimensional system,

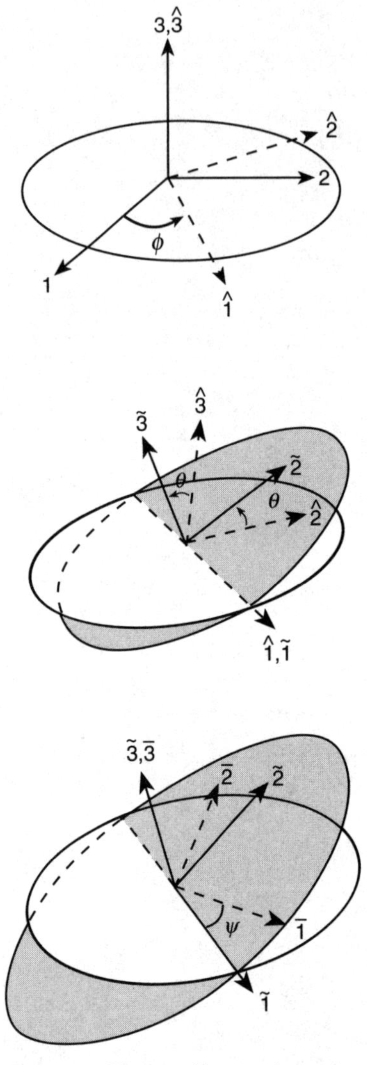

(a) Rotate by ϕ in the 1-2 plane

(b) Rotate by θ in the $\hat{2}$-$\hat{3}$ plane

(c) Rotate by ψ in the $\tilde{1}$-$\tilde{2}$ plane

Figure 1.6 The specific rotations of axes that define the Euler angles θ, ϕ, and ψ.

must be

$$[e_{i\hat{l}}] = \begin{bmatrix} \cos\phi & -\sin\phi & 0 \\ \sin\phi & \cos\phi & 0 \\ 0 & 0 & 1 \end{bmatrix} \tag{1.14}$$

or, in terms of vector components

$$A_i = \sum_{\hat{l}} e_{i\hat{l}} A_{\hat{l}} \qquad i = 1, 2, 3 \quad \hat{l} = \hat{1}, \hat{2}, \hat{3} \tag{1.15}$$

The second rotation, shown in Fig. 1.6*b*, rotates the $(\hat{1}, \hat{2}, \hat{3})$ axes by an angle θ in the $\hat{2}$-$\hat{3}$ plane. We label these new axes as $(\tilde{1}, \tilde{2}, \tilde{3})$ and the transformation of the matrix is

$$[e_{\hat{i}\tilde{k}}] = \begin{bmatrix} 1 & 0 & 0 \\ 0 & \cos\theta & -\sin\theta \\ 0 & \sin\theta & \cos\theta \end{bmatrix} \tag{1.16}$$

or, in terms of vector components

$$A_{\hat{i}} = \sum_{\tilde{k}} e_{\hat{i}\tilde{k}} A_{\tilde{k}} \tag{1.17}$$

We note that by combining Eqs. (1.15) and (1.17), we have

$$A_i = \sum_{\hat{l}} e_{i\hat{l}} \left[\sum_{\tilde{l}} e_{\hat{l}\tilde{k}} A_{\tilde{k}} \right] = \sum_{\hat{l}} \sum_{\tilde{k}} e_{i\hat{l}} e_{\hat{l}\tilde{k}} A_{\tilde{k}} \tag{1.18}$$

or

$$A_i = \sum_{\tilde{k}} e_{i\tilde{k}} A_{\tilde{k}} \tag{1.19}$$

where

$$e_{i\tilde{k}} = \sum_{\hat{l}} e_{i\hat{l}} e_{\hat{l}\tilde{k}} \tag{1.20}$$

Equation (1.19) gives the vector components in the (1, 2, 3) system terms of the vector components in the $(\tilde{1}, \tilde{2}, \tilde{3})$ system.

For our third and final rotation, we rotate the $(\tilde{1}, \tilde{2}, \tilde{3})$ system by an angle ψ in the $\tilde{1}$-$\tilde{2}$ plane, as shown in Fig. 1.6*c*.

We denote the final set of coordinates obtained in this manner as $(\bar{1}, \bar{2}, \bar{3})$, and the transformation matrix is given by

$$[e_{\tilde{k}\bar{j}}] = \begin{bmatrix} \cos\psi & -\sin\psi & 0 \\ \sin\psi & \cos\psi & 0 \\ 0 & 0 & 1 \end{bmatrix} \tag{1.21}$$

or, in terms of vector components

$$A_{\tilde{k}} = \sum_{\bar{j}} e_{\tilde{k}\bar{j}} A_{\bar{j}} \tag{1.22}$$

By combining Eqs. (1.19) and (1.22) we obtain

$$A_i = \sum_{\bar{j}} e_{i\bar{j}} A_{\bar{j}} \tag{1.23}$$

where

$$e_{i\bar{j}} = \sum_{\tilde{k}} e_{i\tilde{k}} e_{\tilde{k}\bar{j}} \tag{1.24}$$

or carrying-out these operations using Eqs. (1.14), (1.16), and (1.21) gives

$$\left[e_{i\bar{j}}\right] = \begin{bmatrix} \cos\phi\cos\psi - \sin\phi\cos\theta\sin\psi & -\cos\phi\sin\psi - \sin\phi\cos\theta\cos\psi & \sin\phi\sin\theta \\ \sin\phi\cos\psi + \cos\phi\cos\theta\sin\psi & -\sin\phi\sin\psi + \cos\phi\cos\theta\cos\psi & -\cos\phi\sin\theta \\ \sin\theta\sin\psi & \sin\theta\cos\psi & \cos\theta \end{bmatrix} \tag{1.25}$$

Equations (1.23) and (1.25) are used extensively in applications, as the examples below show; the $(\bar{1}, \bar{2}, \bar{3})$ or $(\bar{x}, \bar{y}, \bar{z})$ system is taken to be fixed in the body of the molecule and the (1, 2, 3) or (x, y, z) system is taken as the laboratory or space-fixed frame of reference. Then Eqs. (1.23) and (1.25) give the formula for computing the lab-frame components of a vector from the knowledge of the body-frame components.

Example 1.1 Dipole-Dipole Interaction Potential A dipole consists of two point charges at a fixed distance apart. A dipole-dipole interaction system is illustrated in Fig. 1.7. For convenience of notation, we label the charge species as a, b, c, and d. The potential energy of interaction between the dipoles is the sum of the coulombic interactions between the point charges, i.e.,

$$u = \frac{q_a q_c}{r_{ac}} + \frac{q_a q_d}{r_{ad}} + \frac{q_b q_c}{r_{bc}} + \frac{q_b q_d}{r_{bd}} \tag{1.26}$$

where the separation distance vectors are given by (Fig. 1.7)

$$\mathbf{r}_{ac} = \mathbf{r} + \frac{1}{2}\mathbf{b}_2 - \frac{1}{2}\mathbf{b}_1 \tag{1.27}$$

$$\mathbf{r}_{ad} = \mathbf{r} - \frac{1}{2}\mathbf{b}_2 - \frac{1}{2}\mathbf{b}_1 \tag{1.28}$$

$$\mathbf{r}_{bc} = \mathbf{r} + \frac{1}{2}\mathbf{b}_2 + \frac{1}{2}\mathbf{b}_1 \tag{1.29}$$

$$\mathbf{r}_{bd} = \mathbf{r} - \frac{1}{2}\mathbf{b}_2 + \frac{1}{2}\mathbf{b}_1 \tag{1.30}$$

and $r_{ac} \equiv |\mathbf{r}_{ac}|, r_{ad} = |\mathbf{r}_{ad}|$, and so on. In the above equations, $\mathbf{r}$ is the separation distance vector between the *center of mass* of the two dipoles, and $\mathbf{b}_2$ and $\mathbf{b}_1$ are vectors fixed to each dipole (dipole vectors). For simplicity, we will assume that all charge magnitudes are equal and the charge separation distance for each dipole is equal, i.e., $|\mathbf{b}_2| = |\mathbf{b}_1| = b$. Now, with $\mathbf{b}_1 = b\mathbf{e}_{\bar{z}_1}$ and $\mathbf{b}_2 = b\mathbf{e}_{\bar{z}_2}$ the dipole vectors can be expressed in terms of a fixed, laboratory

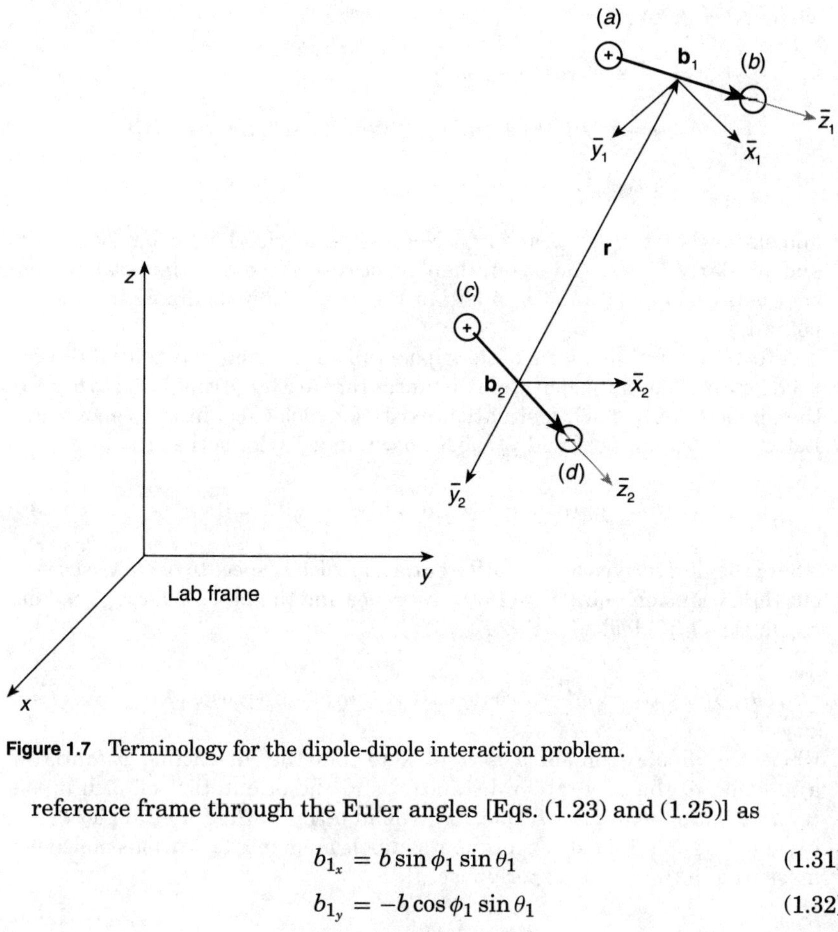

Figure 1.7 Terminology for the dipole-dipole interaction problem.

reference frame through the Euler angles [Eqs. (1.23) and (1.25)] as

$$b_{1_x} = b \sin\phi_1 \sin\theta_1 \tag{1.31}$$

$$b_{1_y} = -b \cos\phi_1 \sin\theta_1 \tag{1.32}$$

$$b_{1_z} = b \cos\theta_1 \tag{1.33}$$

and

$$b_{2_x} = b \sin\phi_2 \sin\theta_2 \tag{1.34}$$

$$b_{2_y} = -b \cos\phi_2 \sin\theta_2 \tag{1.35}$$

$$b_{2_z} = b \cos\theta_2 \tag{1.36}$$

Also, without loss of generality, we can take the separation distance vector to lie along the laboratory z-axis, i.e.,

$$\mathbf{r} = r\,\mathbf{e}_z \tag{1.37}$$

Now, with these expressions, we can readily write the separation distances as

$$r_{ac}^2 = r^2 + (\mathbf{b}_2 - \mathbf{b}_1)\cdot\mathbf{r} + \frac{1}{4}(\mathbf{b}_2 - \mathbf{b}_1)^2 \tag{1.38}$$

where $\mathbf{A}^2 \equiv \mathbf{A} \cdot \mathbf{A}$, or

$$
\begin{aligned}
r_{ac}^2 &= r^2 + br(\cos\theta_2 - \cos\theta_1) \\
&\quad + \frac{1}{4}b^2[\sin^2\theta_2 + \sin^2\theta_1 - 2\sin\theta_1 \sin\theta_2 \cos(\phi_1 - \phi_2) \\
&\quad + (\cos\theta_2 - \cos\theta_1)^2]
\end{aligned}
\tag{1.39}
$$

and similarly for r_{ad}^2, r_{bc}^2, and r_{bd}^2. Now, we can write $1/r_{ac}$ from Eq. (1.39), and similarly $1/r_{ad}$, and so on, then substitute these into the potential energy expression, Eq. (1.26), to obtain the desired dipole-dipole interaction potential.

A further simplified form of the dipole-dipole interaction potential is often used for dipole-dipole separation distances that are large compared to $b \equiv |\mathbf{b}|$, the dipole length. Such a situation exists for molecules in the vapor state. Let $y_{ac} = 1/r_{ac}$ and expand y_{ac} about $b = 0$ in a Taylor series, i.e.,

$$y_{ac}(b) = y_{ac}(b=0) + y'_{ac}(b=0) + \frac{1}{2}y''_{ac}(b=0) + \cdots \tag{1.40}$$

where the derivatives denote differentiation with respect to b. Now, carrying out the expansion using Eq. (1.39) and repeating the process for r_{ad}, r_{bc} and r_{bd} in Eq. (1.26) leads to (Prob. 1.10)

$$u(r, \theta_1, \phi_1, \theta_2, \phi_2) = -\frac{\mu^2}{r^3}[2\cos\theta_1 \cos\theta_2 - \sin\theta_1 \sin\theta_2 \cos(\phi_1 - \phi_2)] \tag{1.41}$$

where the dipole moment $\mu \equiv |q|b$. Note that the interaction potential is a function of the separation distance r and the orientation of each dipole relative to the laboratory frame, as given in terms of the Euler angles θ_1, θ_2, ϕ_1, and ϕ_2. Table 1.2 lists experimental dipole moments for various molecules measured in the gas or vapor phase.

TABLE 1.2 Experimental Dipole Moments (μ) of Various Molecules Measured in the Gas Phase

Substance	Name	μ (in D)
H_2O	Water	1.854
HNO_3	Nitric acid	2.17
H_2O_2	Hydrogen peroxide	1.573
CHN	Hydrogen cyanide	2.984
CH_2O	Formaldehyde	2.332
C_2H_6O	Dimethyl ether	1.30
C_3H_6O	Acetone	2.88
C_7H_8	Toluene	0.375

SOURCE: *Handbook of Chemistry and Physics*, 77th ed., CRC Press, 1996-97. Note that $1D$(Debye) $\equiv 1 \times 10^{-18}$ esu $\cdot$ cm $\equiv 3.33564 \times 10^{-30}$ C $\cdot$ m, where C is Coulomb, the unit of charge.

Interlude 1.2 Water: The Essential Dipole of Life Water, by far is the single most important earthly example of a molecular dipole. Water is a triangular shaped molecule consisting of one oxygen and two hydrogen atoms. The particular electron distribution in a single water molecule gives it an asymmetric charge distribution and, hence, a dipole moment. A simple point charge arrangement with oxygen and hydrogen carrying partial negative and positive charges, respectively, is often used in practice to model the charge distribution.[1] Water has a relatively high dipole moment and consequently a high dielectric constant of about 80, that is, water diminishes the strength of electrostatic interactions between highly charged protein molecules leading to more short-ranged or localized protein-protein interactions. Water also plays an important role in the formation of proteins from their constituent amino acids. Proteins are used by the body as catalysts for reactions, messengers between cells, and many other functions. Naturally occurring or functionally active proteins are globular in construction with charged, hydrophilic molecules on their exteriors and nonpolar, hydrophobic molecules within their interiors (Fig. 1.8). Protein formation in a water

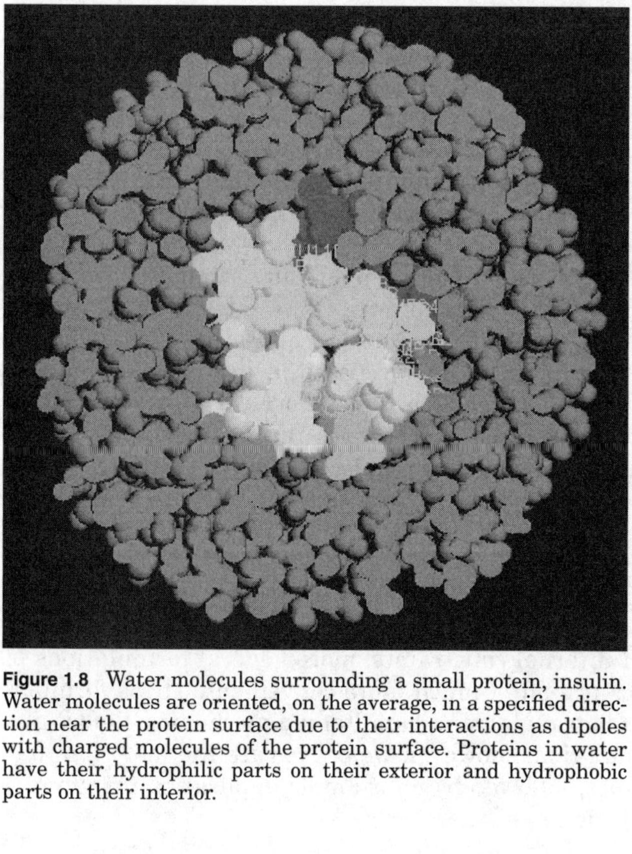

Figure 1.8 Water molecules surrounding a small protein, insulin. Water molecules are oriented, on the average, in a specified direction near the protein surface due to their interactions as dipoles with charged molecules of the protein surface. Proteins in water have their hydrophilic parts on their exterior and hydrophobic parts on their interior.

environment is essential for this configuration. In addition, nonpolar, hydrophobic molecules, such as fats and lipids, tend to cluster when immersed in water. Cell membranes, for example, are highly ordered arrangements of lipid molecules made possible by this hydrophobic effect. In total, about 70 percent of a typical mammalian cell, by weight, is water. It is no wonder that scientists in search of life on other planets first look to see if water is present. Without water, life as we know it would not be possible.

1.3 Introduction to Classical Mechanics

Newton's second law of motion, $\mathbf{F} = d(m\mathbf{v})/dt$, relates the total force $\mathbf{F}$ acting on an object with its time rate of change of momentum $\mathbf{p}$, where $\mathbf{p} = m\mathbf{v}$. The force $\mathbf{F}$ acts through the center of mass of the object. Newton's second law allows us to determine the trajectory of the center of mass of any object relative to a space-fixed (laboratory) or inertial coordinate system. Similarly, the torque acting on an object, $\mathbf{T}_0$, about some reference point "0" is equal to the time rate of change of the angular momentum $\mathbf{L}_0$ of the object about the reference point, i.e., $\mathbf{T}_0 = d\mathbf{L}_0/dt$.

As shown below, the motion of an arbitrarily shaped body, such as a "structured" molecule, can be decomposed in terms of translational motion of the center of mass of the molecule ($\mathbf{F} = d\mathbf{p}/dt$) and rotation of the molecule about its center of mass ($\mathbf{T}_c = d\mathbf{L}_c/dt$). It is to be noted that in writing the force and torque equations, all vector components are with reference to an inertial or space-fixed (laboratory) frame.

Often molecules are represented as spherically symmetric particles with central force interactions, such as the L-J interaction discussed previously. For these systems the torque equation is meaningless. For most systems and applications discussed in this text, we will assume spherically symmetric particles. For more complex particles or structured molecules, we must specify both the intermolecular force, as discussed in Sec.1.2, and the intermolecular torque. For structured molecules, solutions to the torque equation are complex, requiring the introduction of so-called Euler angles or other representations of the various degrees of rotational freedom. Also, there may be molecular constraints that prohibit certain types of motions, for example, "preferred" bond rotations in molecules. For finite-size, structured molecules and for molecules with internal constraints, more general formulations of classical mechanics are often much more convenient. These formulations are known as lagrangian and hamiltonian mechanics. Before discussing these formulations, however, we begin with the more familiar newtonian mechanics, followed by an example application involving a dipole in an electric field.

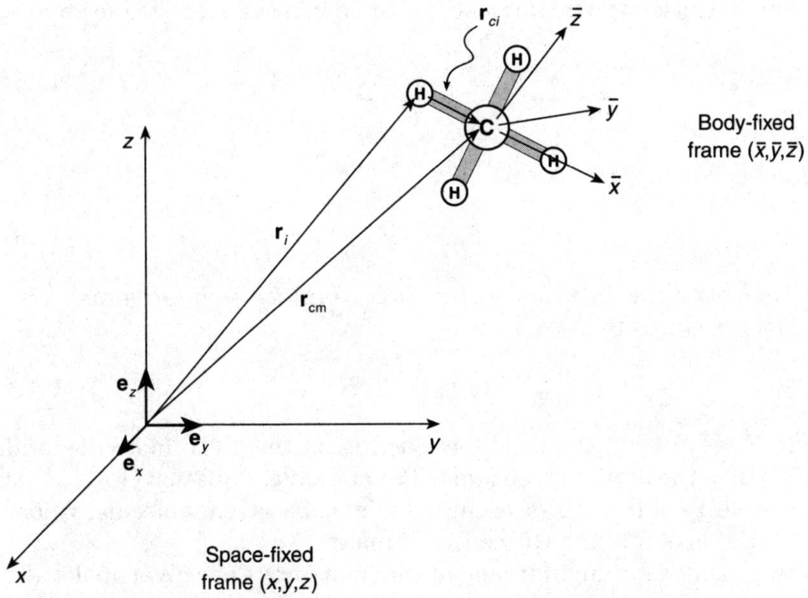

Figure 1.9 Illustration and terminology for the classical mechanics of a structured molecule. (CH_4 is shown here.)

1.3.1 Newtonian mechanics

Consider the methane (CH_4) molecule shown in Fig. 1.9. Each atom (i) of a given CH_4 molecule moves according to Newton's second law of motion

$$\dot{\mathbf{p}}_i = \underbrace{\sum_{\substack{j \\ j \neq i}} \mathbf{F}_{ij}}_{\text{intramolecular}} + \underbrace{\mathbf{F}_i^e}_{\text{intermolecular}} \tag{1.42}$$

where $\mathbf{F}_{ij}$ is the force acting on atom (i) by all other (j) atoms of the molecule (intramolecular force), $\mathbf{F}_i^e$ is the force on atom (i) due to atoms from neighboring molecules (intermolecular force) or other external forces, and $\dot{\mathbf{p}}_i$ is the time rate of change of momentum of atom (i).

Introducing the center of mass vector for a given molecule as

$$\mathbf{r}_{\text{cm}} = \frac{\sum_i m_i \mathbf{r}_i}{\sum_i m_i} = \frac{\sum_i m_i \mathbf{r}_i}{m} \tag{1.43}$$

where m is the total mass of the molecule, and summing Eq. (1.42) over all atoms of the molecule gives

$$\sum_i \dot{\mathbf{p}}_i = m\,\ddot{\mathbf{r}}_{\text{cm}} = \sum_i \sum_{\substack{j \\ j \neq i}} \mathbf{F}_{ij} + \sum_i \mathbf{F}_i^e \tag{1.44}$$

Now by rewriting the intramolecular force in a symmetrized form as

$$\sum_i \sum_{\substack{j \\ j \neq i}} \mathbf{F}_{ij} = \frac{1}{2} \sum_i \sum_{\substack{j \\ j \neq i}} (\mathbf{F}_{ij} + \mathbf{F}_{ji}) \tag{1.45}$$

and using Newton's third law

$$\mathbf{F}_{ij} = -\mathbf{F}_{ji} \tag{1.46}$$

the sum over the intramolecular force term is seen to vanish and Eq. (1.44) reduces to

$$\dot{\mathbf{p}} = m\,\ddot{\mathbf{r}}_{\mathrm{cm}} = \mathbf{F} \tag{1.47}$$

where $\mathbf{F} \equiv \sum_i \mathbf{F}_i^e$ is the total force acting on the given molecule, and $\mathbf{p} = \sum_i \mathbf{p}_i$ is the total momentum of the molecule. Equation (1.47) gives the trajectory of the center of mass (point) of a given molecule, where the force $\mathbf{F}$ acts through the center of mass.

Now consider the angular momentum and torque of a given molecule. Taking the cross product of the vector $\mathbf{r}_i$ and Eq. (1.42) and summing over all atoms gives

$$\sum_i \mathbf{r}_i \times \dot{\mathbf{p}}_i = \sum_i \mathbf{r}_i \times \sum_{\substack{j \\ j \neq i}} \mathbf{F}_{ij} + \sum_i \mathbf{r}_i \times \mathbf{F}_i^e \tag{1.48}$$

or,

$$\sum_i \mathbf{r}_i \times \dot{\mathbf{p}}_i = \sum_i \sum_{\substack{j \\ j \neq i}} \mathbf{r}_i \times \mathbf{F}_{ij} + \sum \mathbf{r}_i \times \mathbf{F}_i^e \tag{1.49}$$

Now, we define the total angular momentum vector $\mathbf{L}$ as

$$\mathbf{L} \equiv \sum_i \mathbf{r}_i \times \mathbf{p}_i \tag{1.50}$$

and, thus,

$$\dot{\mathbf{L}} = \sum_i \dot{\mathbf{r}}_i \times \mathbf{p}_i + \sum_i \mathbf{r}_i \times \dot{\mathbf{p}}_i = \sum_i \mathbf{r}_i \times \dot{\mathbf{p}}_i \tag{1.51}$$

Since $\dot{\mathbf{r}}_i$ and $\mathbf{p}_i$ are colinear vectors, their cross product must vanish.

It will be shown below to be more convenient to reexpress the angular momentum in terms of center of mass ($\mathbf{r}_{\mathrm{cm}}$) and internal coordinates ($\mathbf{r}_{c_i}$) as (Fig. 1.9)

$$\mathbf{r}_i = \mathbf{r}_{\mathrm{cm}} + \mathbf{r}_{c_i} \tag{1.52}$$

Differentiating with respect to time, we also have

$$\mathbf{v}_i = \mathbf{v} + \mathbf{v}_{c_i} \tag{1.53}$$

The angular momentum can thus be rewritten as

$$\begin{aligned}\mathbf{L} &= \sum_i (\mathbf{r}_{\text{cm}} + \mathbf{r}_{c_i}) \times m_i(\mathbf{v} + \mathbf{v}_{c_i}) \\ &= \sum_i m_i \mathbf{r}_{\text{cm}} \times \mathbf{v} + \sum_i m_i \mathbf{r}_{\text{cm}} \times \mathbf{v}_{c_i} \\ &\quad + \sum_i m_i \mathbf{r}_{c_i} \times \mathbf{v} + \sum_i m_i \mathbf{r}_{c_i} \times \mathbf{v}_{c_i}\end{aligned} \tag{1.54}$$

Now, the two middle terms in this last expression vanish by virtue of the properties†

$$\sum_i m_i \mathbf{r}_{c_i} = 0 \tag{1.55}$$

and

$$\sum_i m_i \mathbf{v}_{c_i} = 0 \tag{1.56}$$

Thus, from Eq. (1.54) we can write

$$\begin{aligned}\mathbf{L} &= \sum_i m_i \mathbf{r}_{\text{cm}} \times \mathbf{v} + \sum_i m_i \mathbf{r}_{c_i} \times \mathbf{v}_{c_i} \\ &= m\,\mathbf{r}_{\text{cm}} \times \mathbf{v} + \sum_i m_i \mathbf{r}_{c_i} \times \mathbf{v}_{c_i} \\ &= \mathbf{L}_0 + \mathbf{L}_c\end{aligned} \tag{1.57}$$

where

$$\mathbf{L}_0 \equiv m\,\mathbf{r}_{\text{cm}} \times \mathbf{v} \tag{1.58}$$

is the angular momentum of the center of mass about the origin, and

$$\mathbf{L}_c = \sum_i m_i \mathbf{r}_{c_i} \times \mathbf{v}_{c_i} \tag{1.59}$$

is the angular momentum of the molecule about its center of mass. Note that all vectors are with reference to the space-fixed or laboratory

†To see these write $\sum_i m_i \mathbf{r}_i = \sum_i m_i(\mathbf{r}_{\text{cm}} + \mathbf{r}_{c_i})$. But, by definition $m\mathbf{r}_{\text{cm}} = \sum_i m_i \mathbf{r}_i$, so $\sum_i m_i \mathbf{r}_{c_i} = 0$. Differentiating with respect to time gives $\sum_i m_i \mathbf{v}_{c_i} = 0$.

frame, such as

$$\mathbf{r}_{c_i} = x_{c_i}\mathbf{e}_x + y_{c_i}\mathbf{e}_y + z_{c_i}\mathbf{e}_z \tag{1.60}$$

Returning to Eq. (1.49) we have

$$\dot{\mathbf{L}} = \dot{\mathbf{L}}_0 + \dot{\mathbf{L}}_c = \sum_i \sum_{\substack{j \\ j\neq i}} \mathbf{r}_i \times \mathbf{F}_{ij} + \sum_i \mathbf{r}_i \times \mathbf{F}_i^e \tag{1.61}$$

Writing the first term on the right-hand side of the above equation as [cf. Eq. (1.45)]

$$\sum_i \sum_{\substack{j \\ j\neq i}} \mathbf{r}_i \times \mathbf{F}_{ij} = \frac{1}{2} \sum_i \sum_{\substack{j \\ j\neq i}} (\mathbf{r}_i \times \mathbf{F}_{ij} + \mathbf{r}_j \times \mathbf{F}_{ji}) \tag{1.62}$$

and using Newton's third law gives

$$\sum_i \sum_{\substack{j \\ j\neq i}} \mathbf{r}_i \times \mathbf{F}_{ij} = \frac{1}{2} \sum_i \sum_{\substack{j \\ j\neq i}} (\mathbf{r}_i - \mathbf{r}_j) \times \mathbf{F}_{ij} \tag{1.63}$$

Thus, this term vanishes when the interatomic force lies along $(\mathbf{r}_i - \mathbf{r}_j)$, which is the usual case. Thus, Eq. (1.61) becomes

$$\begin{aligned}\dot{\mathbf{L}} = \dot{\mathbf{L}}_0 + \dot{\mathbf{L}}_c &= \sum_i (\mathbf{r}_{\text{cm}} + \mathbf{r}_{c_i}) \times \mathbf{F}_i^e \\ &= \mathbf{r}_{\text{cm}} \times \mathbf{F} + \sum_i \mathbf{r}_{c_i} \times \mathbf{F}_i^e \end{aligned} \tag{1.64}$$

Now, using Eq. (1.47) for the center of mass motion, the above equation reduces to

$$\dot{\mathbf{L}}_c = \mathbf{T}_c \tag{1.65}$$

where

$$\dot{\mathbf{L}}_c \equiv \sum_i m_i \mathbf{r}_{c_i} \times \ddot{\mathbf{r}}_{c_i} \tag{1.66}$$

from Eq. (1.59), and

$$\mathbf{T}_c \equiv \sum_i \mathbf{r}_{c_i} \times \mathbf{F}_i^e \tag{1.67}$$

is the external torque acting on the molecule about its center of mass. Equations (1.47) and (1.65) describe the translation motion of the center of mass of the molecule and the rotation of the molecule about its center of mass, respectively.

For rigid molecules, the angular momentum $\mathbf{L}_c$ can be expressed in terms of the moment of inertia tensor $\mathbf{I}$ and the angular velocity ω of the molecule relative to the laboratory frame. The advantage of using the moment of inertia tensor lies in the fact that any symmetry of the body can be exploited to simplify the development of the component forms of Eq. (1.65). These approaches are treated extensively in texts on classical mechanics, such as those listed at the end of this chapter, and will not be further considered here. In the example below, we simply take the "brute-force" approach and work with Eqs. (1.65) through (1.67) directly.

In any event, the angular momentum (or angular velocity and moment of inertia) and torque can be expressed in terms of the Euler angles that give the orientation of the molecule relative to the laboratory frame. The resulting "orientational" equations of motion can then be solved in terms of the Euler angles giving the time dependency of molecular rotations as shown in Example 1.2. Extensions to pseudorigid molecules are also possible, which have important applications in molecular spectroscopy.[2] Usually, however, it is not a simple matter to express the torque in terms of the Euler angles and alternative methods (lagrangian and hamiltonian) are preferred, as discussed more fully in the following sections. In these alternative methods, one begins with the specification of the total energy (kinetic plus potential) of the molecule, which is usually easier to write, rather than beginning with the forces and torques.

Example 1.2 Newtonian Dynamics of a Dipole in an External Electric Field
Consider a rigid dipole consisting of two equal and opposite charges at a fixed separation distance b, as sketched in Fig. 1.10. We wish to determine the equations of motion for the dipole in a uniform electric field $\mathbf{E}_0$. The force

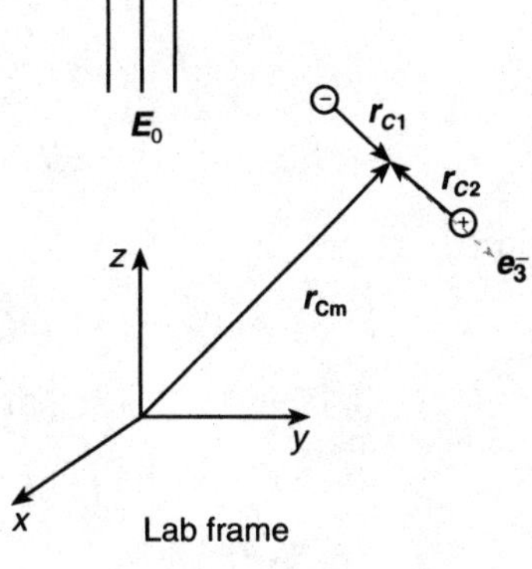

Figure 1.10 A single dipole in a uniform electric field.

on each charge is thus

$$\mathbf{F}_1^e = -|q|\mathbf{E}_0 \tag{1.68}$$

$$\mathbf{F}_2^e = +|q|\mathbf{E}_0 \tag{1.69}$$

Note that the net force acting on the dipole is zero, i.e., there will be no translational motion of the center of mass due to the $\mathbf{E}_0$-field. The torque about the center of mass acting on the dipole is given from Eq. (1.67) as

$$\begin{aligned}\mathbf{T}_c &= \sum_{i=1}^{2} \mathbf{r}_{c_i} \times \mathbf{F}_i^e \\ &= |q|(\mathbf{r}_{c2} - \mathbf{r}_{c1}) \times \mathbf{E}_0\end{aligned} \tag{1.70}$$

Now,

$$\mathbf{r}_{c1} = +(b/2)\mathbf{e}_{\bar{3}} \tag{1.71}$$

$$\mathbf{r}_{c2} = -(b/2)\mathbf{e}_{\bar{3}} \tag{1.72}$$

and

$$\mathbf{r}_{c1} - \mathbf{r}_{c2} = b\mathbf{e}_{\bar{3}} \tag{1.73}$$

Using Eq. (1.25) we can transform the unit vector $\mathbf{e}_{\bar{3}}$ in the body-fixed frame to the laboratory frame leading to

$$\mathbf{r}_{c1} - \mathbf{r}_{c2} = b[\sin\phi \sin\theta \mathbf{e}_1 - \cos\phi \sin\theta \mathbf{e}_2 + \cos\theta \mathbf{e}_3] \tag{1.74}$$

Now with $\mathbf{E}_0 = E_0\mathbf{e}_3$ we can easily carry out the cross products to obtain the torque acting on the dipole as

$$\mathbf{T}_c = b|q|E_0[\sin\phi \sin\theta \mathbf{e}_2 - \cos\phi \sin\theta \mathbf{e}_1] \tag{1.75}$$

In order to write the angular momentum $\mathbf{L}_c$ we need the components of the vectors $\mathbf{r}_{c1}$ and $\mathbf{r}_{c2}$, which again follow from the transformations of Eqs. (1.71) and (1.72) as

$$x_{c1} = (b/2)\sin\phi \sin\theta \tag{1.76}$$

$$x_{c2} = -x_{c1} \tag{1.77}$$

$$y_{c1} = -(b/2)\cos\phi \sin\theta \tag{1.78}$$

$$y_{c2} = -y_{c1} \tag{1.79}$$

$$z_{c1} = (b/2)\cos\theta \tag{1.80}$$

$$z_{c2} = -z_{c1} \tag{1.81}$$

and $r_{c1}^2 = r_{c2}^2 = b^2/4$. We also need the time derivatives

$$\dot{x}_{c1} = (b/2)[\sin\phi\cos\theta\dot{\theta} + \cos\phi\sin\theta\dot{\phi}] \tag{1.82}$$

$$\dot{x}_{c2} = -\dot{x}_{c1} \tag{1.83}$$

$$\dot{y}_{c1} = -(b/2)[\cos\phi\cos\theta\dot{\theta} - \sin\phi\sin\theta\dot{\phi}] \tag{1.84}$$

$$\dot{y}_{c2} = -\dot{y}_{c1} \tag{1.85}$$

$$\dot{z}_{c1} = -(b/2)\sin\theta\dot{\theta} \tag{1.86}$$

$$\dot{z}_{c2} = -\dot{z}_{c1} \tag{1.87}$$

Now, from the definition of the angular momentum, Eq. (1.59), we have, for example, the x-component

$$\begin{aligned} L_{cx} &= m\,(y_{c1}\dot{z}_{c1} - z_{c1}\dot{y}_{c1}) + m(y_{c2}\dot{z}_{c2} - z_{c2}\dot{y}_{c2}) \\ &\frac{mb^2}{2}[\cos\phi\dot{\theta} - \sin\theta\cos\theta\sin\phi\dot{\phi}] \end{aligned} \tag{1.88}$$

Thus, the x-component of the rotational equation of motion, Eq. (1.65), is

$$\frac{mb^2}{2}\frac{d}{dt}[\cos\phi\dot{\theta} - \sin\theta\cos\theta\sin\phi\dot{\phi}] = -b|q|E_0\cos\phi\sin\theta \tag{1.89}$$

Similar equations can be derived for the y and z components, leading to the three simultaneous equations for the functions $\theta(t)$, $\phi(t)$, and $\psi(t)$. Thus, we have an initial value problem for determining the time rate of charge of the Euler angles that describe the orientation of the dipole in the **E**-field. We also note that at equilibrium, the net torque on the dipole must be zero. From Eq. (1.75) this results in $\theta = 0, \pi$ and from Fig. (1.10) it is evident that $\mathbf{e}_{\bar{3}} = \mathbf{e}_3$ or $-\mathbf{e}_3$, that is the dipole is aligned with the z-axis (with the **E**-field).

1.4 Lagrangian Mechanics

A more general formulation of the mechanics of particle systems is based on Hamilton's principle, or the principle of least action. This principle states that the action S defined as

$$S = \int_{t_1}^{t_2} L(q, \dot{q}, t)\,dt \tag{1.90}$$

takes on the least possible value, where L is the lagrangian, equal to the total kinetic energy (K) minus the total potential energy (Φ), $L = K - \Phi$, q [$q \equiv (q_1, q_2, \ldots)$] are called the generalized coordinates and $\dot{q}$[$\dot{q} \equiv (\dot{q}_1, \dot{q}_2, \ldots)$] are called the generalized velocities. The generalized coordinates q represent the mechanical degrees of freedom of the particle(s) motion. For example, in single rigid body motion, the generalized coordinates are simply the x, y, and z coordinates of the body center of mass and the three Euler angles (θ, ϕ, ψ) describing the body's orientation relative to a space-fixed reference frame. For most systems, the total kinetic energy K can be written as a quadratic function of the generalized velocities, i.e.,

$$K = \sum_i a_i(q)\dot{q_i}^2 \tag{1.91}$$

Using a variational formulation, it can be shown that Hamilton's principle of least action leads to the following Lagrange's equations of motion†

$$\frac{d}{dt}\left(\frac{\partial L}{\partial \dot{q}_i}\right) - \frac{\partial L}{\partial q_i} = 0 \qquad (\text{i} = 1, 2, \ldots, s) \tag{1.92}$$

where s represents the number of degrees of freedom or generalized coordinates in the system (e.g., six for the rigid body). The beauty and practicality of Lagrange's formulation is that any (holonomic) constraints‡ can be directly introduced in the formulation by appropriately assigning the degrees of freedom, as Example 1.3 illustrates. Note that Lagrange's equations represent s, second-order differential equations.

Example 1.3 Application of Lagrangian Mechanics Consider a simple two-dimensional pendulum. An object of mass M is maintained at a distance R from a fixed point by a massless rigid rod. The motion of the system is restricted to a vertical plane containing the rod. The single degree of freedom is the polar angle θ ($s = 1$) giving the angular position of the rod relative to a space-fixed frame of reference.

We first must formulate the lagrangian L in terms of θ and $\dot{\theta}$:

$$L = K - \Phi = \frac{1}{2}m(\dot{x}^2 + \dot{y}^2) + mgx \tag{1.93}$$

Using $x = R\cos\theta$, $y = R\sin\theta$, $\dot{x} = -R\sin\theta\dot{\theta}$, and $\dot{y} = R\cos\theta\dot{\theta}$, we obtain

$$L = \frac{1}{2}mR^2\dot{\theta}^2 + mgR\cos\theta \tag{1.94}$$

Lagrange's equation of motion for the single degree of freedom is

$$\frac{d}{dt}\left(\frac{\partial L}{\partial \dot{\theta}}\right) = \frac{\partial L}{\partial \theta} \tag{1.95}$$

†See Landau and Lifshitz, 1980 in the Further Reading section at the end of this chapter.

‡Holonomic constraints are those that do *not* depend on the generalized velocities.

Substituting for the lagrangian gives

$$mR^2 \frac{d}{dt}(\dot{\theta}) = mgR \sin\theta \tag{1.96}$$

or

$$\frac{d^2\theta}{dt^2} = \frac{g}{R} \sin\theta \tag{1.97}$$

The solution to this second-order equation is

$$\theta = -\frac{g}{R} \sin\theta + C_1 t + C_2 \tag{1.98}$$

where C_1 and C_2 are integration constants ($C_1 = d\theta/dt$), specified by the initial position and velocity.

Note how, in general, Lagrange's equations of motion lead to s, second-order equations, where s is the number of generalized coordinates. Also, to appreciate the straightforwardness of lagrangian mechanics, try repeating this example using newtonian mechanics. Homework problem 1.9 repeats Example 1.2 for the dipole-electric field system using lagrangian mechanics.

1.5 Hamilton's Equations of Motion

Another useful form of the equations of motion, in addition to newtonian and lagrangian forms, are Hamilton's equations of motion. To see these, we first define the so-called conjugate momentum p_i as

$$p_i = \frac{\partial L}{\partial \dot{q}_i} \qquad (i = 1, 2, \ldots, s) \tag{1.99}$$

Next, we define the hamiltonian as the total energy of the system, kinetic and potential, i.e.,†

$$H(p, q, t) = \sum_i p_i \dot{q}_i - L(q, \dot{q}, t) \tag{1.100}$$

where it is understood that the $\dot{q}_i$ on the right-hand side of Eq. (1.100) are to be eliminated in favor of the p_i using Eq. (1.99). Thus, we have

$$dH = \sum_i p_i d\dot{q}_i + \sum_i \dot{q}_i dp_i - dL \tag{1.101}$$

†Using the general expression for K, we have

$$H = \sum_i p_i \dot{q}_i - \sum_i a_i \dot{q}_i^2 + \Phi$$

and from Eq. (1.94), $p_i = 2a_i \dot{q}_i$; thus, $H = K + \Phi$. Also note that $q \equiv (q_1, q_2, \ldots, q_s)$, $p \equiv (p_1, p_2, \ldots, p_s)$, and so on.

Now, using the chain rule and assuming that L does not depend explicitly on time, i.e., L depends on t only implicitly through the dependence of q and $\dot{q}$ on t (see the previous example), we have

$$dL = \sum_i \frac{\partial L}{\partial q_i} dq_i + \sum_i \frac{\partial L}{\partial \dot{q}_i} d\dot{q}_i \tag{1.102}$$

Using Eqs. (1.95) and (1.102) in Eq. (1.101) gives

$$dH = -\sum_i \dot{p}_i dq_i + \sum_i \dot{q}_i dp_i \tag{1.103}$$

Finally, using the chain rule again for $H = H(p, q)$ where H, like L, is assumed not to depend explicitly on time, we have

$$dH = \sum_i \frac{\partial H}{\partial q_i} dq_i + \sum_i \frac{\partial H}{\partial p_i} dp_i \tag{1.104}$$

Comparing Eqs. (1.103) and (1.104) gives

$$\dot{q}_i = \frac{\partial H}{\partial p_i} \tag{1.105}$$

and

$$\dot{p}_i = -\frac{\partial H}{\partial q_i} \tag{1.106}$$

Equations (1.105) and (1.106) are known as Hamilton's equations of motion and constitute 2s first-order differential equations for the 2s unknown functions $p_i(t)$ and $q_i(t)$. Hamilton's equations of motion are extremely useful in statistical mechanical formulations as will be demonstrated in the next chapter. Because we have assumed no explicit time dependence of the lagrangian or hamiltonian, we have from Eqs. (1.104) to (1.106) that $dH/dt = 0$, that is, the total energy H is conserved, or is a constant of motion, for such systems. In general, the time integration of Eqs. (1.105) and (1.106) will result in motion along a constant total energy surface. We note, however, that it can be shown that Hamilton's equations also hold when the lagrangian or hamiltonian exhibit explicit time dependency ("nonconservative" systems).[3]

We conclude this section with one final note on Hamilton's equations. As shown in Prob. 1.5, Lagrange's equation of motion retains its form under any transformation from the coordinates $q_1, q_2, \ldots$ to any other independent set $Q_1, Q_2, \ldots$, where $Q_i = Q_i(q_1, q_2, \ldots, t)$. However, Hamilton's equations do not necessarily retain their form under any type of transformation $Q_i = Q_i(p_1, q_1, p_2, q_2, \ldots, t)$ and

$P_i = P_i(p_1, q_1, p_2, q_2, \ldots, t)$. Those transformations that do retain the form of Hamilton's equations are said to be "*canonical transformations*." In practice, canonical transformations can often greatly simplify the solutions to Hamilton's equations. For more discussion on the use of canonical transformations in theoretical mechanics, please see the Further Reading section list at the end of the chapter.

1.6 Summary

Our goals in this chapter have been to introduce the general ideas and importance of statistical mechanics and the need to study equilibrium and nonequilibrium states together (and not separately under different subjects or approaches). Nonequilibrium systems often represent perturbations from equilibrium states, and there is a set of thermodynamic variables, which will be introduced formally in Chap. 4, common to both equilibrium and nonequilibrium systems and analyses. In (global) nonequilibrium states, these variables depend on space and/or time, whereas in (global) equilibrium states there are no such dependencies. In general, thermodynamic variables represent "averages" over many millions of molecules or atoms of the system. Averages or mean values are formulated from probability functions; so the goal of statistical mechanics is to predict that these probability functions, which we will demonstrate explicitly later, depend on specific intermolecular interactions and dynamics of the system.

To accomplish our goals and to prepare us for the remaining chapters of this text, a general understanding of intermolecular forces and interactions, as well as a basic understanding of classical mechanics is needed. The intermolecular interaction forces include electrostatic, van der Waals, and Born repulsive forces. These intermolecular forces can vary greatly among different substances, and they can lead to differences in their phase state existence as a gas, liquid, or solid, as well as differences in their physical properties, such as viscosity and conductivity. Classical mechanics was described under three complimentary approaches—newtonian, lagrangian, and hamiltonian mechanics. Lagrangian and hamiltonian mechanics are the more general approaches and they contain the more familiar newtonian mechanics as a special case. Lagrangian and hamiltonian mechanics are based on first identifying the spatial degrees of freedom for any particular system. For example, a simple pendulum swinging in a plane has one degree of freedom, namely the polar angle describing the orientation of the pendulum with respect to some reference axis. As we will see in the next chapter, Hamilton's mechanics formulation has direct application in the formulation of the Liouville equation, which is the basic equation or starting point of statistical mechanics.

Problems

1.1 Show that Newton's second law of motion for two interacting particles can be expressed by

$$\frac{d^2\mathbf{r}_{cm}}{dt^2} = 0$$

and

$$M\frac{d^2\mathbf{r}}{dt^2} = \mathbf{F}(\mathbf{r})$$

where $\mathbf{r}_{cm}$ is the center of mass vector

$$\mathbf{r}_{cm} \equiv \frac{m_1\mathbf{r}_1 + m_2\mathbf{r}_2}{m_1 + m_2}$$

$\mathbf{r} = \mathbf{r}_1 - \mathbf{r}_2$, and M is called the reduced mass,

$$M \equiv \frac{m_1 m_2}{m_1 + m_2}$$

1.2 *Simple Harmonic Oscillation in Solids*: A pair of interacting atoms or molecules in the solid state are generally found near the minimum in the interaction potential, or

$$F(r_0) = -\frac{du(r_0)}{dr} = 0$$

where r_0 is the separation distance at the minimum. Any interaction force can be linearized by the following two-term Taylor series expansion

$$F(r) = F(r_0) + (r - r_0)\frac{dF}{dr}|_{r=r_0} + \cdots$$

or using the definition of the maximum

$$F(r) \cong A(r - r_0)$$

where $A = \frac{dF}{dr}|_{r=r_o}$ is a constant for this linearized restoring force. Derive the constant A for the following potentials:

1. Mie potential
2. Lennard-Jones potential
3. Coulombic potential

1.3 A realistic potential function for an ionic system (gas, liquid, or solid) includes a Born repulsive term in addition to the coulombic term as

$$u(r) = \frac{B}{r^n} - \frac{e^2}{4\pi r\epsilon_0}$$

Show a sketch of this potential function and express the constant B in terms of the minimum in the potential at interatomic separation distance r_0, defined by $F(r_0) = 0$.

1.4 Using a Taylor series expansion about r_0 derive an analytical expression for the restoring force for the potential given in Prob. 1.3.

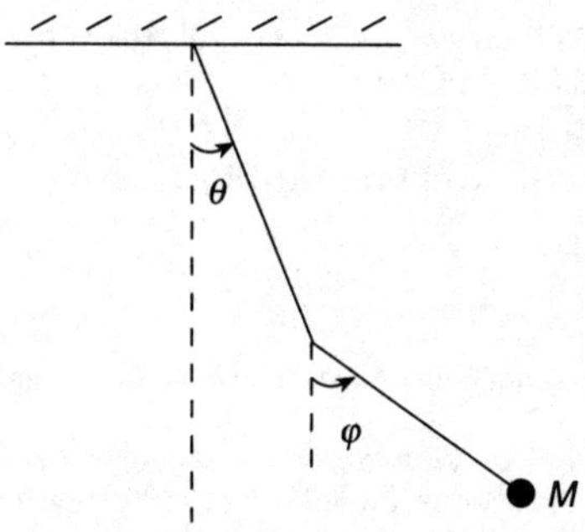

Figure 1.11 The double, planar pendulum.

1.5 Lagrange's equations of motion have the remarkable and useful property that they have the same form in any coordinate system. Prove this by writing $x = x(q_1, q_2, q_3)$, $y = y(q_1, q_2, q_3)$, and $z = z(q_1, q_2, q_3)$, then transforming Eq. (1.92) from $L = L(q_1, q_2, q_3)$ to $L = L(x, y, z)$. *Hint*: Express L in terms of Φ and K and use Eq. (1.91).

1.6 For a single point molecule with lagrangian

$$L = \frac{1}{2}m(\dot{x}^2 + \dot{y}^2 + \dot{z}^2) - u(x, y, z)$$

show that Lagrange's equations, Eq. (1.92), reduce to Newton's equations of motion.

1.7 *The Double Pendulum*: An object of mass M is connected to two massless rods of length R_1 and R_2 as shown in the figure below (Fig. 1.11). The motion of the system is restricted to the vertical plane containing the two rods. By writing the lagrangian in terms of the two angles θ and ϕ, determine the equations of motion for the object.

1.8 Derive the y and z components of the rotational equation of motion in Example 1.2.

1.9 Repeat Example 1.2 using lagrangian mechanics. Note that the kinetic energy of the dipole is

$$T = \frac{1}{2}m(\mathbf{r}_1 \cdot \mathbf{r}_1) + \frac{1}{2}m(\mathbf{r}_2 \cdot \mathbf{r}_2)$$

This can be rewritten using

$$\mathbf{r}_1 = \mathbf{r}_{\text{cm}} + \mathbf{r}_{c1}$$

$$\mathbf{r}_2 = \mathbf{r}_{\text{cm}} + \mathbf{r}_{c2} = \mathbf{r}_{\text{cm}} - \mathbf{r}_{c1}$$

Substituting,

$$T = m(\mathbf{r}_{\text{cm}} \cdot \mathbf{r}_{\text{cm}}) + m(\mathbf{r}_{c1} \cdot \mathbf{r}_{c1})$$

The first term is the kinetic energy of the center-of-mass of the dipole, and the second term is the rotational kinetic energy of the dipole. The generalized coordinates in this case are $(\mathbf{r}_{\mathrm{cm}}, \theta, \psi, \phi)$. The potential energy of the dipole can be found in Eq. (2.53) in Chap. 2.

1.10 Use the Taylor series expansion, Eq. (1.40), to obtain Eq. (1.41).

References

1. K. Toukan and A. Rahman, *Molecular dynamics study of atomic motions in water*, *Phys. Rev. B*, **31**, 2643–2648, 1985.
2. E.B. Wilson, Jr., J.C: Decius, and P.C. Cross, *Molecular Vibrations. The Theory of Infrared and Raman Vibrational Spectra*, Dover Publications, New York, 1980. (In particular, see Chap. 11, "The Separation of Rotation and Vibration," which develops the hamiltonian for pseudorigid molecules.)
3. R. C. Tolman, *The Principles of Statistical Mechanics*, Dover Publications, New York, 1979.

Further Reading

L. D. Landau and E. M. Lifshitz, *Mechanics, Course of Theoretical Physics*, 3rd ed. Vol.1, Pergamons, New York, 1988. [(An excellent treatment of Lagrangian mechanics is given in this text.) Other excellent theoretical mechanics texts include H. Goldstein, *Classical Mechanics*, 2nd ed., Addison-Wesley, MA, 1980. E.A. Desloge, *Classical Mechanics*, Vols. I and II, Krieger Publishing, FL, 1989.]

M. Rigby, E. B. Smith, W. A. Wakeham, and G. C. Maitland, *The Forces Between Molecules*, Oxford Science Publications, New York, 1986. [An excellent introduction to the subject of intermolecular interaction forces.]

D. Tabor, *Gases, Liquids and Solids*, Penguin, Baltimore, MD, 1969.

A. J. Walton, *Three Phases of Matter*, Oxford Science Publications, New York, 1983. [Both Tabor's book and Walton's book provide comprehensive introductory material on the three phases of matter.]

Chapter

2

Phase Space and the Liouville Equation

"Now, I'll just have to start
To be twice as careful and be twice as smart.
I'll watch out for trouble in front and
back sections
By aiming my eyeballs in different directions"

—THEODOR SEUSS GEISEL,
I Had Trouble Getting to Solla Sollew

2.1 Introduction: The Merger of Classical Mechanics and Probability Theory

As we noted in the first chapter, a complete description of the millions of interacting molecules or atoms in any given system is a hopelessly complex undertaking. Fortunately, such a task is not at all necessary since it will be shown that the observed behavior and properties of the system can be adequately obtained from a study of the interactions of a much smaller number of molecules or atoms. Most of the remainder of this text is dedicated to demonstrating exactly how this is done. Generally speaking, our observations and property measurements of a given system are really averages in some sense over space, time, or numbers of atoms. For example, we cannot instantly know the simultaneous positions and velocities of all atoms of a system by any known measuring device. Because of the finite measurement time or discrete size of the device we can only hope to measure either some kind of temporal or spatial integrated behavior. This is not a bad thing as we can still, for all practical purposes, uniquely characterize the system from this average observed behavior.

Now, it should be clear that we have introduced the word "average" in our discussion, and certainly we have used it rather loosely. Averages imply variations, and variations are formally described in terms of probability, or more precisely, probability functions. In short, we would like to ask ourselves what is the *probability* of finding the atoms or molecules of the system with various positions and velocities at any given time since it is the probability that can be formally used to calculate the observed average behavior.

The merger of probability theory and classical mechanics is accomplished by the so-called Liouville equation, which is considered to be the fundamental equation of statistical mechanics. From this equation we can obtain a comprehensive description of both the equilibrium and nonequilibrium behavior of matter. In this chapter, we will derive the famous Liouville equation from a simple differential *mass balance* approach. In this case, the mass will represent a system of points in a multidimensional space. Each point contains all the information about the system at a particular time. The Liouville equation obtained here will be called upon in each of the subsequent chapters in our quest to describe the observed behavior and properties of any particular system.

2.2 Derivation of the Liouville Equation

For simplicity, we first develop the Liouville equation in cartesian coordinates using newtonian mechanics and *spherically symmetric* molecules. Later in this chapter, we will give the corresponding form in generalized coordinates and conjugate momenta. Consider a system of N molecules at time t, as shown in Fig. 2.1. Each molecule (i) is characterized by a position vector $\mathbf{r}_i = x_i\mathbf{e}_x + y_i\mathbf{e}_y + z_i\mathbf{e}_z$, where $\mathbf{e}_x$, $\mathbf{e}_y$, and $\mathbf{e}_z$ are unit vectors in the x, y, and z directions, respectively, and momentum $\mathbf{p}_i = m_i\mathbf{v}_i = m_i d\mathbf{r}_i/dt(\mathbf{p}_i = p_{xi}\mathbf{e}_x + p_{yi}\mathbf{e}_y + p_{zi}\mathbf{e}_z)$. It is convenient to think of the $3N$ position coordinates and the $3N$ momentum coordinates for the system of N molecules as one single point in a $6N$-dimensional space, as depicted in Fig. 2.1. The $6N$ coordinates are considered to be the independent coordinates in the system. This single point is called a *phase point* and the $6N$-dimensional space is termed *phase space*. Each molecule is assumed to move according to classical mechanics, i.e., Newton's second law of motion is applicable to any ith molecule as

$$\mathbf{F}_i = \frac{d}{dt}\left(m_i \frac{d\mathbf{r}_i}{dt}\right) \tag{2.1}$$

or, for constant mass, m_i

$$\mathbf{F}_i = \frac{d\mathbf{p}_i}{dt} \tag{2.2}$$

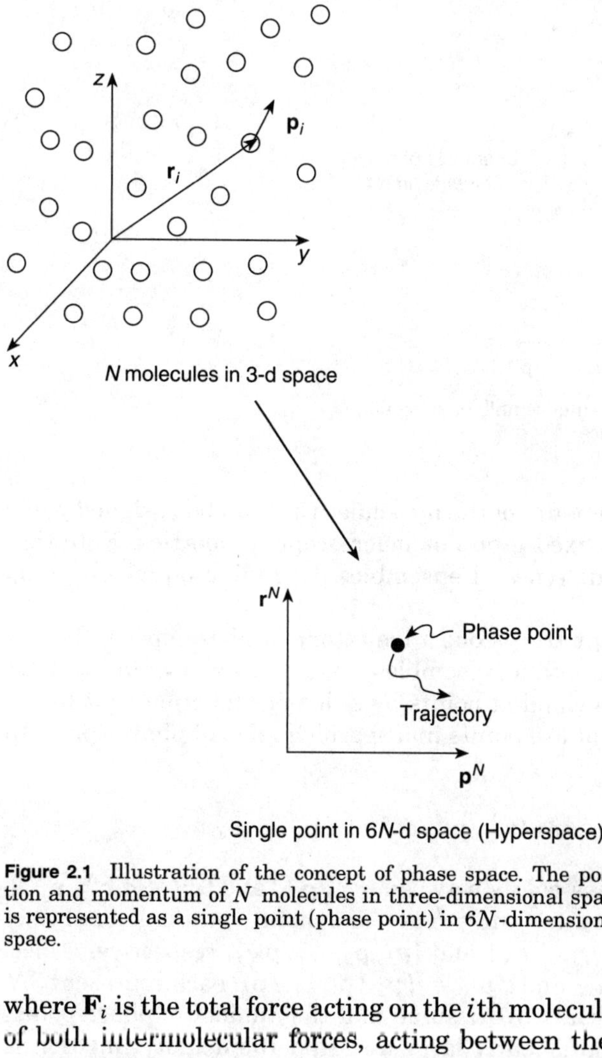

Figure 2.1 Illustration of the concept of phase space. The position and momentum of N molecules in three-dimensional space is represented as a single point (phase point) in $6N$-dimensional space.

where $\mathbf{F}_i$ is the total force acting on the ith molecule. This force consists of both intermolecular forces, acting between the molecules, as well as any external forces that may exist. The intermolecular force, for example, might be as simple as a coulombic electrostatic force between charged molecules, and the external force could be a gravitational force or an external electric field force acting on charged molecules.

As each of the N molecules moves through space according to Eq. (2.2), the phase point traces out a unique trajectory in phase space.

Next consider a collection, or ensemble of phase points. Each phase point of the ensemble represents the same group of molecules, with certain identical groups or macroscopic properties (such as number density and total energy), differing only in their *individual* positions and momenta. In short, there are many different values or "realizations" of

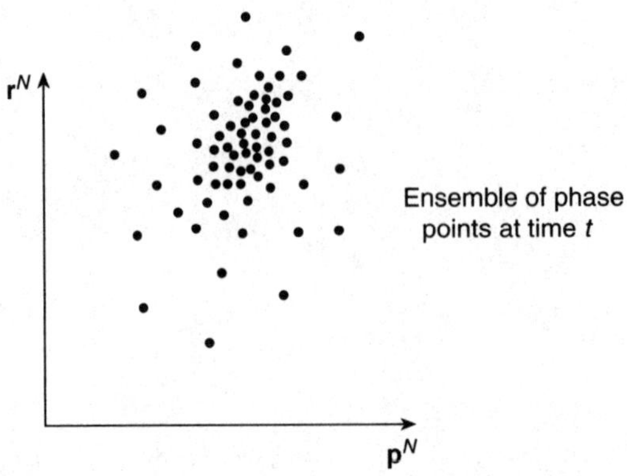

Figure 2.2 Illustration of an ensemble or collection of phase points in phase space.

the positions and momenta of the molecules that can be assigned while maintaining certain fixed group or macroscopic properties. Note that there can be different types of ensembles depending on which group properties are fixed.

As each phase point traces out a trajectory in phase space, the collection of points, at any time, resembles a "cloud" as shown in Fig. 2.2.[†] We can describe this cloud of points by a density function that represents the number of phase points in a specific region of phase space. In particular,

$$\rho_N(\mathbf{r}^N, \mathbf{p}^N, t)d\mathbf{r}^N d\mathbf{p}^N$$

is the number of phase points between $(\mathbf{r}^N, \mathbf{p}^N)$ and $(\mathbf{r}^N + d\mathbf{r}^N, \mathbf{p}^N + d\mathbf{p}^N)$ at time t, where $\mathbf{r}^N$ and $\mathbf{p}^N$ are shorthand notations for the set of coordinates $\{\mathbf{r}_1, \mathbf{r}_2, \ldots, \mathbf{r}_N\}$ and $\{\mathbf{p}_1, \mathbf{p}_2, \ldots, \mathbf{p}_N\}$, respectively; also, $d\mathbf{r}^N \equiv d\mathbf{r}_1 d\mathbf{r}_2 \ldots d\mathbf{r}_N$ and $d\mathbf{p}^N \equiv d\mathbf{p}_1 d\mathbf{p}_2 \ldots d\mathbf{p}_N$ each represent $3N$ dimensional differentials. We assume that the number of phase points in the ensemble is sufficiently large and that the density function is continuous.

Since ρ_N is a density function, it obeys a conservation equation similar in development to the equation of continuity, or mass conservation equation, in fluid mechanics. The Liouville equation simply represents a conservation equation for the phase points through a fixed differential volume of phase space, called a *hypercube*. The rate of phase points into the hypercube through the faces at x_1 is the flux $\rho_N \dot{x}_1$ times the

[†]Note that the collection of *trajectories* of phase points would appear like a "bowl of spaghetti."

(hyperspace) cross-sectional area at x_1, i.e.,

$$\rho_N \dot{x}_1(\Delta y_1 \Delta z_1 \ldots \Delta z_N \Delta p_{1x} \ldots \Delta p_{Nz})|_{x_1}$$

and similarly for the other faces at y_1 through z_N. The rate of phase points into the hypercube through the face at p_{1x} is the flux $\rho_N \dot{p}_{1x}$ times the (hyperspace) cross-sectional area at p_{1x}, i.e.,

$$\rho_N \dot{p}_{1x}(\Delta x_1 \ldots \Delta z_N \Delta p_{1y} \Delta p_{1z} \ldots \Delta p_{Nz})|_{p_{1x}}$$

and similarly for the other faces at p_{1y} through p_{Nz}. The rate of phase points leaving the hypercube through the face at $x_1 + \Delta x_1$ is

$$\rho_N \dot{x}_1(\Delta y_1 \Delta z_1 \ldots \Delta z_N \Delta p_{1x} \ldots \Delta p_{Nz})|_{x_1+\Delta x_1}$$

and similarly for the other faces at $y_1 + \Delta y_1$ through $z_N + \Delta z_N$. The rate of phase points leaving the hypercube through the face at $p_{1x} + \Delta p_{1x}$ is

$$\rho_N \dot{p}_{1x}(\Delta x_1 \ldots \Delta z_N \Delta p_{1y} \Delta p_{1z} \ldots \Delta p_{Nz})|_{p_{1x}+\Delta p_{1x}}$$

and similarly for the other faces at $p_{1y} + \Delta p_{1y}$ through $p_{Nz} + \Delta p_{Nz}$.

Since phase points are not created or destroyed, the total rate of phase points entering the hypercube minus the number of phase points leaving the hypercube must equal the total number of phase points accumulating in the hypercube, or

$$\begin{aligned}
&\rho_n \dot{x}_1(\Delta y_1 \Delta z_1 \ldots \Delta z_n \Delta p_{1x} \ldots \Delta p_{Nz})|_{x_1} - \rho_n \dot{x}_1(\Delta y_1 \Delta z_1 \ldots \Delta z_n \Delta p_{1x} \ldots \Delta p_{Nz})|_{x_1+\Delta x_1} \\
&\qquad + \cdots \qquad - \cdots \\
&+\rho_n \dot{p}_{1x}(\Delta x_1 \ldots \Delta z_n \Delta p_{1y} \Delta p_{1z} \ldots \Delta p_{Nz})|_{p_{1x}} - \rho_n \dot{p}_{1x}(\Delta x_1 \ldots \Delta z_n \Delta p_{1y} \Delta p_{1z} \ldots \Delta p_{Nz})|_{p_{1x}+\Delta p_{1x}} \\
&\qquad + \cdots \qquad - \cdots \\
&= \frac{\partial \rho_n}{\partial t}(\Delta x_1 \ldots \Delta z_n \Delta p_{1x} \ldots \Delta p_{Nz})
\end{aligned} \tag{2.3}$$

Dividing through by the volume of the hypercube, $(\Delta x_1 \ldots \Delta p_{Nz})$, and taking limits as Δx_1 through $\Delta p_{Nz} \to 0$ leads to the Liouville equation

$$-\sum_{i=1}^{3N} \left[\frac{\partial}{\partial x_i}(\rho_N \dot{x}_i) + \frac{\partial}{\partial p_i}(\rho_N \dot{p}_i) \right] = \frac{\partial \rho_N}{\partial t} \tag{2.4}$$

where we have let $(x_1, x_2, x_3, x_4, \ldots)$ represent $(x_1, y_1, z_1, x_2, \ldots)$ and $(p_1, p_2, p_3, p_4, \ldots)$ represent $(p_{1x}, p_{1y}, p_{1z}, p_{2x}, \ldots)$.

Alternatively, in vector notation, the Liouville equation reads

$$\frac{\partial \rho_N}{\partial t} = -\sum_{i=1}^{N} \left[\frac{\partial}{\partial \mathbf{r}_i} \cdot (\dot{\mathbf{r}}_i \rho_N) + \frac{\partial}{\partial \mathbf{p}_i} \cdot (\dot{\mathbf{p}}_i \rho_N) \right] \tag{2.5}$$

Now, we can substitute Newton's second law of motion, Eq. (2.2), and the relationship $\dot{\mathbf{r}}_i = \mathbf{p}_i/m_i$ into Eq. (2.5) above to obtain

$$\frac{\partial \rho_N}{\partial t} = -\sum_{i=1}^{N} \left[\frac{\mathbf{p}_i}{m_i} \cdot \frac{\partial \rho_N}{\partial \mathbf{r}_i} + \mathbf{F}_i(\mathbf{r}^N) \cdot \frac{\partial \rho_N}{\partial \mathbf{p}_i} \right] \tag{2.6}$$

where we have assumed that the force acting on particle i is only a function of the position coordinates $\mathbf{r}^N \equiv (\mathbf{r}_1, \mathbf{r}_2, \ldots, \mathbf{r}_N)$.†

As will be shown in subsequent chapters, the solution to the Liouville equation for the N-particle density function ρ_N is the basis for determination of the equilibrium and nonequilibrium properties of matter. However, because of the large number of dimensions ($6N$), solutions to the Liouville equation represent formidable problems! Fortunately, as will be shown in Chaps. 4 and 5, the equilibrium and nonequilibrium properties of matter can usually be expressed in terms of lower-ordered or "reduced" density functions. For example, the thermodynamic and transport properties of dilute gases can be expressed in terms of the two-molecule density function $\rho_2(\mathbf{r}_1, \mathbf{r}_2, \mathbf{p}_1, \mathbf{p}_2, t)$. In Chap. 3 we will examine the particular forms of the reduced Liouville equation.

To conclude this section with a more in-depth analysis, we note that in the development of the Liouville equation it has been tacitly assumed that the total particle number density, N, is a fixed constant. In this sense, the total N-particle system is considered to be "isolated." However, it is possible to consider a particular subsystem or system with, say, $(N - M)$ fixed particles and a "surroundings" with M fixed particles constituting the total N-particle system or "universe." Then, to describe the subsystem interacting through contact with its surroundings, a reduced form of the Liouville equation in the $(N - M)$ space is needed. In this description, however, we are still restricted to only force-type interactions between the system and surroundings, and particle exchange is not allowed. Homework Prob. 4.13 given in Chap. 4 examines the form of the Liouville equation when the particle number varies or when particle exchange is allowed. In the subject of transport phenomena, which will be given later in Chaps. 5 and 6, the macroscopic transport equations, such as the equations of fluid mechanics and heat transfer, are derived from the Liouville equation, assuming an isolated N-particle system without any regard for the surroundings. Then, the interactions between the system and surroundings are treated through proposed macroscopic boundary conditions—for example, the familiar *no-slip* boundary conditions in fluid mechanics or momentum transport phenomena. In general, the macroscopic boundary conditions may

---

†A velocity dependent force (magnetic force) is considered in Prob. 2.2.

be determined from experimental observations, as in the no-slip condition, or from separately derived conditions based on a local, simplified analysis of the reduced Liouville equation described earlier.

Interlude 2.1 The Abstract Concept of Hyperspace We have introduced *hyperspace* as a $6N$ multidimensional space. The independent coordinates are the $3N$ position variables and the $3N$ momentum variables for the N total molecules in the system. It is impossible to draw such a system in three dimensions, so we must think of hyperspace in abstract or mathematical terms.

As a simple example of this abstraction, consider the equation for a circle (two-dimensional) and a sphere (three-dimensional), respectively,

$$x_1^2 + x_2^2 = R^2$$
$$x_1^2 + x_2^2 + x_3^2 = R^2$$

Mathematically, both equations give the collection of points an equal distance R from the origin. For an N-dimensional *hypersphere*, we would express the collection of points an equal distance R from the origin as

$$x_1^2 + x_2^2 + \cdots + x_N^2 = R^2$$

Let's try to compute the (hyper) volume of this N-dimensional hypersphere. First, look at the following integral

$$\int_{-\infty}^{\infty} \cdots \int_{-\infty}^{\infty} e^{-(x_1^2+x_2^2+\cdots+x_N^2)} dx_1 dx_2 \cdots dx_N$$
$$= \left[\int_{-\infty}^{\infty} e^{-t^2} dt\right]^N = (\pi)^{N/2}$$

Now, noting that a differential spherical volume in three dimensions is $4\pi R^2 dR$ (where $R \in [0, \infty]$), in N dimensions we write the differential volume as $S_N R^{N-1} dR$, where S_N remains to be specified.

The integral above can thus be alternatively written as

$$\int_0^{\infty} e^{-R^2}(R^{N-1}S_N)dR = S_N \int_0^{\infty} e^{-R^2} R^{N-1} dR$$
$$= \frac{1}{2} S_N \left(\frac{N}{2} - 1\right)!$$

where $(R^{N-1}S_N)$ denotes the surface area of the N-dimensional hypersphere of radius R. Comparing the results of the two integrations, we have

$$S_N = \frac{2\pi^{N/2}}{(\frac{N}{2} - 1)!}$$

So the volume of the hypersphere V_{hs} is

$$V_{hs} = \int_0^R S_N R'^{N-1} dR' = \frac{\pi^{N/2}}{(N/2)!} R^N$$

where R' is a "dummy" variable of integration. Note that for $N = 3$, $\left(\frac{3}{2}\right)! = \left(\frac{3}{4}\right) \pi^{1/2}$; thus, $V_{hs} = \frac{4}{3}\pi R^3$ as expected.†

2.3 The Liouville Equation in Terms of Generalized Coordinates and Conjugate Momenta

As explained in Chap. 1, it is often advantageous to work with generalized coordinates $q = \{q_1, q_2, \ldots, q_s\}$ and the conjugate momenta $p = \{p_1, p_2, \ldots, p_s\}$ for more complex molecular systems such as systems with finite-sized molecules and special constraints. Introducing a density function ρ_s in this new space

$$\rho_s(q, p, t)\, dq\, dp$$

as the number of phase points between (q, p) and $(q + dq, p + dp)$, we can follow the exact same steps as in Sec. 2.1 to obtain

$$\frac{\partial \rho_s}{\partial t} = -\sum_{i=1}^{s} \left[\frac{\partial(\rho_s \dot{q}_i)}{\partial q_i} + \frac{\partial(\rho_s \dot{p}_i)}{\partial p_i} \right] \tag{2.7}$$

where s is the total number of degrees of freedom.

Substituting Hamilton's equations of motion, Eqs. (1.104) and (1.105), into Eq. (2.7) and simplifying gives (Prob. 2.5)

$$\frac{\partial \rho_s}{\partial t} = -\sum_{i=1}^{s} \left[\frac{\partial H}{\partial p_i} \frac{\partial \rho_s}{\partial q_i} - \frac{\partial H}{\partial q_i} \frac{\partial \rho_s}{\partial p_i} \right] \tag{2.8}$$

Equation (2.8) is the desired Liouville equation written in terms of the generalized coordinates and conjugate momenta.

Now, we can show an important property of the generalized Liouville equation. If we write

$$\rho_s = \rho_s(q_i, p_i, t)$$

according to the chain rule for partial derivatives

$$\frac{d\rho_s}{dt} = \frac{\partial \rho_s}{\partial t} + \sum_{i=1}^{s} \left[\frac{\partial \rho_s}{\partial q_i} \dot{q}_i + \frac{\partial \rho_s}{\partial p_i} \dot{p}_i \right] \tag{2.9}$$

---

†For manipulation of factorial fractions see Ref. 1.

Comparing Eqs. (2.8) and (2.9) yields

$$\frac{d\rho_s}{dt} = 0 \tag{2.10}$$

Thus, the density of phase points behaves as an "incompressible fluid." In other words, a differential volume element in phase space containing a certain number of phase points can change its shape, but not size.

Finally, before leaving this section, we note another important aspect of the Liouville equation regarding transformation of phase space variables. We noted in Chap. 1 that Hamilton's equations of motion retain their form only for so-called *canonical transformations*. Consequently, the form of the Liouville equation given above is also invariant to only canonical transformations. Furthermore, it can be shown that the Jacobian for canonical transformations is unity, i.e., there is no expansion or contraction of a phase space volume element in going from one set of phase space coordinates to another. A simple example of a single particle in three dimensions can be used to effectively illustrate this point.† Considering, for example, two representations, viz., cartesian and spherical coordinates and their associated conjugate momenta, we have

$$dx\,dy\,dz\,dp_x\,dp_y\,dp_z = dr\,d\theta\,d\phi\,dp_r\,dp_\theta\,dp_\phi$$

where p_r, p_θ, and p_ϕ are the momenta conjugate to r, θ, and ϕ, respectively. The simple volume transformations in ordinary space, $dx\,dy\,dz = r^2 \sin\theta\,d\theta\,d\phi\,dr$, do not have any connection to phase space transformations.

Interlude 2.2 Mathematical Essentials: Solution to Linear, First-Order Partial Differential Equations Linear, first-order partial differential equations appear in a number of subjects, including Euler's equations of motion in fluid mechanics, single particle Liouville equation in statistical mechanics, unsteady-state convective flow of heat and mass in transport phenomena, and many others. Many partial differential equation (PDE) texts treat this subject very lightly and usually without regard to applications, including the satisfaction of specific initial or boundary conditions. In order to illustrate the general strategy, consider the following linear PDE

$$f(x, y)u_x + g(x, y)u_y = h(x, y) \tag{2.11}$$

where $u = u(x, y)$, $u_x = \partial u/\partial x$, $u_y = \partial u/\partial y$, and f, g, and h are arbitrary functions of x and y. Now, we introduce a parameter s, where $u = u[s(x, y)]$

†See D. A. McQuarrie, p. 123, in the Further Reading section at the end of this chapter.

and from the chain rule

$$\frac{du}{ds} = \frac{\partial u}{\partial x}\frac{dx}{ds} + \frac{\partial u}{\partial y}\frac{dy}{ds} \tag{2.12}$$

Comparison of Eqs. (2.11) and (2.12) gives

$$\frac{dx}{ds} = f(x, y) \tag{2.13}$$

$$\frac{dy}{ds} = g(x, y) \tag{2.14}$$

$$\frac{du}{ds} = h(x, y) \tag{2.15}$$

Equations (2.13) and (2.14) define a parametric representation of an (x, y) plane curve called a *characteristic* curve. The solution to Eq. (2.15) is then sought along the characteristic curves. Let's see specifically how the procedure works via an example:

$$u_x + 2xu_y = y \tag{2.16}$$

subject to the initial values (initial curve)

$$u(0, y) = 1 + y^2 \qquad 1 < y < 2 \tag{2.17}$$

Thus, we have from Eq. (2.13)

$$x = s + C_1 \tag{2.18}$$

and, without loss of generality, we can take the parameter s to be zero at $x = 0$ giving $C_1 = 0$. Next, from Eq. (2.14)

$$\frac{dy}{ds} = 2x = 2s \tag{2.19}$$

or

$$y = s^2 + C_2 \tag{2.20}$$

The characteristic curves, from Eqs. (2.18) and (2.20) are seen to be parabolas defined by

$$y = x^2 + C_2 \tag{2.21}$$

Now, using Eq. (2.20), the solution to Eq. (2.15) along the characteristic curve is

$$u = \frac{1}{3}s^3 + C_2 s + C_3 \tag{2.22}$$

or, in terms of x,

$$u = \frac{1}{3}x^3 + C_2 x + C_3 \tag{2.23}$$

By substitution of Eq. (2.23) into Eq. (2.16), it can be seen that Eq. (2.23) is a solution to our original PDE only along the curve $y = x^2 + C_2$.

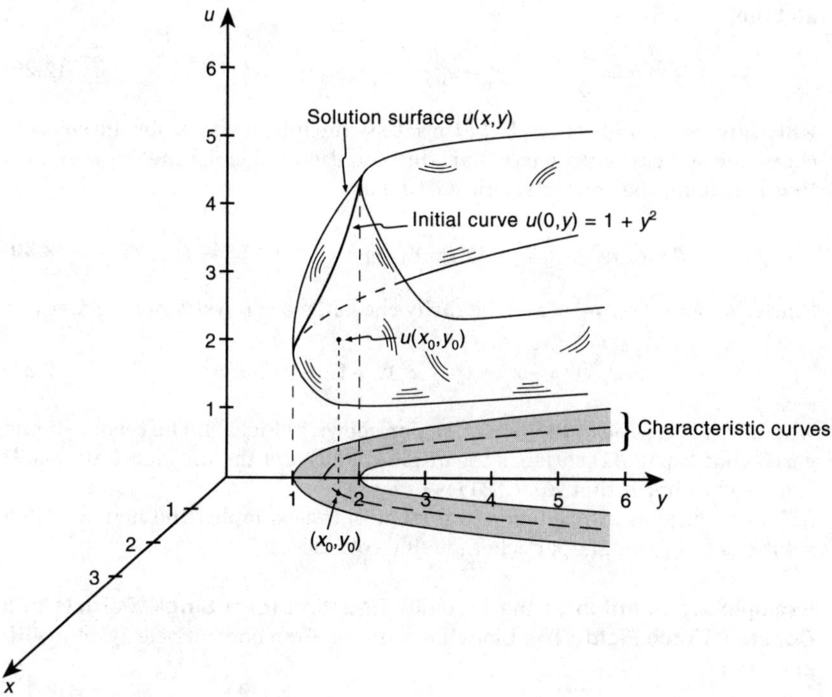

Figure 2.3 Illustration of the solution to a linear, first-order PDE along the characteristic curves.

In order to determine the constants C_2 and C_3 we must link Eqs. (2.23) to the initial values. Let's choose an arbitrary point (x_0, y_0) in the shaded region shown in Fig. 2.3. The characteristic curve through that point is

$$y = x^2 + \left(y_0 - x_0^2\right) \tag{2.24}$$

and the values of u for this curve are

$$u = \frac{1}{3}x^3 + \left(y_0 - x_0^2\right)x + C_3 \tag{2.25}$$

For all values $x = 0$, this gives

$$u = C_3 \tag{2.26}$$

Likewise, the values of $u(0, y) = 1 + y^2$ $(1 < y < 2)$ for values of y along the characteristic curve with $x = 0$ are from Eq. (2.24)

$$u(0, y) = 1 + \left(y_0 - x_0^2\right)^2 \tag{2.27}$$

Comparison of Eqs. (2.26) and (2.27) gives

$$C_3 = 1 + \left(y_0 - x_0^2\right)^2 \tag{2.28}$$

and thus

$$u = \frac{1}{3}x^3 + (y_0 - x_0^2)x + \left[1 + (y_0 - x_0^2)^2\right] \tag{2.29}$$

which gives the values of u that also satisfy the initial values for the selected characteristic curve. In particular, this equation holds for the point (x_0, y_0) that lies along the characteristic curve, i.e.,

$$u(x_0, y_0) = \frac{1}{3}x_0^3 + (y_0 - x_0^2)x_0 + \left[1 + (y_0 - x_0^2)^2\right] \tag{2.30}$$

Since the point (x_0, y_0) was arbitrarily chosen, we can write more generally

$$u(x, y) = \frac{1}{3}x^3 + (y - x^2)x + [1 + (y - x^2)^2] \tag{2.31}$$

where (x, y) must lie in the shaded region shown in Fig. 2.3. The reader should verify that Eq. (2.31) satisfies the original PDE and the initial conditions. It can also be shown that Eq. (2.31) is unique.†

This technique will be followed further in the examples and homework on solutions to the single-particle Liouville equation.

Example 2.1 Solution to the Liouville Equation for a Single Particle in a Constant Force Field The Liouville equation for a one particle system with mass m is

$$\frac{\partial \rho_1}{\partial t} = -\frac{\mathbf{p}_1}{m} \cdot \frac{\partial \rho_1}{\partial \mathbf{r}_1} - \mathbf{F}_1 \cdot \frac{\partial \rho_1}{\partial \mathbf{p}_1} \tag{2.32}$$

where the force $\mathbf{F}_1$ is assumed to be constant.

We seek the solution for a point source, i.e.,

$$\rho_1 = M\delta(\mathbf{r}_1 - \mathbf{r}_1')\delta(\mathbf{p}_1 - \mathbf{p}_1') \qquad \text{at} \quad t = 0 \tag{2.33}$$

where M = total ensemble number
δ = Dirac delta function
$\mathbf{r}_1'$ = initial position
$\mathbf{p}_1'$ = initial momentum.

Note that the Dirac delta function is defined by its properties

$$\delta(x - x_0) = \begin{cases} 0 & x \neq x_0 \\ \infty & x = x_0 \end{cases} \tag{2.34}$$

and

$$\int_{-\infty}^{\infty} \delta(x - x_0)dx = 1 \tag{2.35}$$

†For further treatment, see Ref. 2. The example here was adapted from Sec. 6.1, Ref. 2.

Also, note that we adopt the notation

$$\delta(\mathbf{r}-\mathbf{r}') \equiv \delta(x-x')\delta(y-y')\delta(z-z') \tag{2.36}$$

Following mathematical essentials above, we parameterize our problem as $s = s(t, \mathbf{r}, \mathbf{p})$ with the characteristic curves given according to Eq. (2.11) as

$$t = s \tag{2.37}$$

$$\mathbf{p}_1 = \mathbf{F}_1 s + \mathbf{C}_2 \tag{2.38}$$

$$\mathbf{r}_1 = \frac{1}{2m}\mathbf{F}_1 s^2 + \frac{\mathbf{C}_2}{m}s + \mathbf{C}_3 \tag{2.39}$$

or

$$\mathbf{p}_1 = \mathbf{F}_1 t + \mathbf{C}_2 \tag{2.40}$$

$$\mathbf{r} = \frac{1}{2m}\mathbf{F}_1 t^2 + \frac{\mathbf{C}_2}{m}t + \mathbf{C}_3 \tag{2.41}$$

The characteristic curves, Eqs. (2.40) and (2.41), are simply the trajectories of the particle subject to the constant force $\mathbf{F}_1$. The solution to the density function along the characteristic curves is

$$\rho_1 = C_4 \tag{2.42}$$

Now, we choose an arbitrary point $(\mathbf{r}_{1_0}, \mathbf{p}_{1_0}, t_0)$, and the trajectories (characteristic curves) through that point from Eqs. (2.40) and (2.41) are as follows:

$$\mathbf{p}_1 = \mathbf{F}_1 t + (\mathbf{p}_{1_0} - \mathbf{F}_1 t_0) \tag{2.43}$$

$$\mathbf{r}_1 = \frac{1}{2m}\mathbf{F}_1 t^2 + \frac{t}{m}(\mathbf{p}_{1_0} - \mathbf{F}_1 t_0) + \left[\mathbf{r}_{1_0} + \frac{1}{2m}\mathbf{F}_1 t_0^2 - \frac{t_0}{m}\mathbf{p}_{1_0}\right] \tag{2.44}$$

For our initial condition, Eq. (2.33), the values of $\mathbf{p}_1$ and $\mathbf{r}_1$ along the characteristic curve and for $t = 0$ are obtained from Eqs. (2.43) and (2.44) leading to

$$\rho(0, \mathbf{r}_1, \mathbf{p}_1) = M\delta[(\mathbf{p}_{1_0} - \mathbf{F}_1 t_0) - \mathbf{p}_1']\delta\left[\left(\mathbf{r}_{1_0} + \frac{1}{2m}\mathbf{F}_1 t_0^2 - \frac{t_0}{m}\mathbf{p}_{1_0}\right) - \mathbf{r}_1'\right] \tag{2.45}$$

Comparison to Eq. (2.42) gives

$$\rho = C_4 = M\delta[(\mathbf{p}_{1_0} - \mathbf{F}_1 t_0) - \mathbf{p}_1']\delta\left[\left(\mathbf{r}_{1_0} + \frac{1}{2m}\mathbf{F}_1 t_0^2 - \frac{t_0}{m}\mathbf{p}_{1_0}\right) - \mathbf{r}_1'\right] \tag{2.46}$$

Since the point $(\mathbf{r}_{1_0}, \mathbf{p}_{1_0}, t_0)$ was chosen arbitrarily, we can write the solution as

$$\rho(\mathbf{r}_1, \mathbf{p}_1, t) = M\delta[\mathbf{p}_1 - (\mathbf{p}_1' + \mathbf{F}_1 t)]\delta\left[\mathbf{r}_1 - \left(\mathbf{r}_1' + \frac{t}{m}\mathbf{p}_1 - \frac{1}{2m}\mathbf{F}_1 t^2\right)\right] \tag{2.47}$$

Thus, the initial source follows a simple trajectory dictated by the constant force $\mathbf{F}_1$

$$\mathbf{r}_1(t) = \mathbf{r}_1' + \frac{\mathbf{p}_1}{m}t - \frac{\mathbf{F}_1 t^2}{2m} = \mathbf{r}_1' + \frac{\mathbf{p}_1'}{m}t + \frac{\mathbf{F}_1 t^2}{2m} \tag{2.48}$$

$$\mathbf{p}_1(t) = \mathbf{p}_1' + \mathbf{F}_1 t \tag{2.49}$$

The above example illustrates the development and solution to the single-particle Liouville equation in cartesian coordinates. As discussed in Chap. 1, for structured particles, it is often more fruitful to work in generalized coordinates. Let's revisit the problem of a dipole in an external electric field, Example 1.1, and develop the generalized coordinates, conjugate momenta, hamiltonian, and associated Liouville equation for this system.

Example 2.2 Liouville Equation for a Single Dipole in a Uniform External Electric Field The first step in using lagrangian or hamiltonian mechanics is to write the potential and kinetic energy in terms of the generalized coordinates, which for our dipole problem are the cartesian coordinates of the center of mass of the dipole and the Euler angles describing the orientation of the dipole relative to the lab frame. Again, we take the external field to be oriented with the z-axis,

$$\mathbf{E}_0 = E_0 \mathbf{e}_z \tag{2.50}$$

so that the total potential energy of the dipole is [cf. Eqs. (1.68) and (1.69)]

$$\Phi = \Phi_1 + \Phi_2 = |q|\mathbf{E}_0 z_1 - |q|\mathbf{E}_0 z_2 \tag{2.51}$$

In terms of center of mass and relative coordinates, Eq. (1.52),

$$\Phi = |q|\mathbf{E}_0(z_{c_2} - z_{c_1}) \tag{2.52}$$

and from Eqs. (1.38) and (1.39)

$$\Phi = |q|\mathbf{E}_0 b \cos\theta \tag{2.53}$$

which is the potential energy in terms of the Euler angle θ.

Turning now to the kinetic energy, defined by

$$T = \frac{1}{2}m(\dot{\mathbf{r}}_1 \cdot \dot{\mathbf{r}}_1) + \frac{1}{2}m(\dot{\mathbf{r}}_2 \cdot \dot{\mathbf{r}}_2) \tag{2.54}$$

or, in terms of center of mass and relative coordinates, with $\dot{\mathbf{r}}_{c_2} = -\dot{\mathbf{r}}_{c_1}$ for the dipole,

$$T \ + m(\dot{\mathbf{r}}_{\text{cm}} \cdot \dot{\mathbf{r}}_{\text{cm}}) + m(\dot{\mathbf{r}}_{c_1} \cdot \dot{\mathbf{r}}_{c_1}) \tag{2.55}$$

Using Eqs. (1.34), (1.36), and (1.38), the kinetic energy can be written in terms of the Euler angles as

$$T = m(\dot{x}^2 + \dot{y}^2 + \dot{z}^2) + \frac{mb^2}{4}(\dot{\theta}^2 + \sin^2\theta\dot{\phi}^2) \tag{2.56}$$

where (x, y, z) are the cartesian coordinates of the center of mass of the dipole.

Now, we must eliminate the generalized velocities in terms of the conjugate momenta to write the hamiltonian as required by Eq. (2.51), the Liouville equation. Our second step, therefore, is to obtain expressions for the conjugate momenta from Eq. (1.99). Writing the lagrangian from its definition

$$L = T - \Phi \tag{2.57}$$

and with the generalized coordinates for the dipole given by $q = (x, y, z, \theta, \phi)$, we obtain from Eq. (1.99)

$$p_x = \frac{\partial L}{\partial \dot{x}} = 2m\dot{x} \tag{2.58}$$

$$p_y = \frac{\partial L}{\partial \dot{y}} = 2m\dot{y} \tag{2.59}$$

$$p_z = \frac{\partial L}{\partial \dot{z}} = 2m\dot{z} \tag{2.60}$$

$$p_\theta = \frac{\partial L}{\partial \dot{\theta}} = \frac{mb^2}{2}\dot{\theta} \tag{2.61}$$

and

$$p_\phi = \frac{\partial L}{\partial \dot{\phi}} = \frac{mb^2}{2}\sin^2\theta\dot{\phi} \tag{2.62}$$

Using these relationships, the kinetic energy can be reexpressed from Eq. (2.56) as

$$T = \frac{1}{4m}\left(p_x^2 + p_y^2 + p_z^2\right) + \frac{1}{mb^2}\left(p_\theta^2 + \frac{1}{\sin^2\theta}p_\phi^2\right) \tag{2.63}$$

From the hamiltonian $H = T + \Phi$, we now obtain

$$\frac{\partial H}{\partial p_i} = \frac{1}{m}p_i \qquad i = x, y, z \tag{2.64}$$

$$\frac{\partial H}{\partial x} = \frac{\partial H}{\partial y} = \frac{\partial H}{\partial z} = 0 \tag{2.65}$$

$$\frac{\partial H}{\partial p_\theta} = \frac{2}{mb^2}p_\theta \tag{2.66}$$

$$\frac{\partial H}{\partial p_\phi} = \frac{2}{mb^2\sin^2\theta}p_\phi \tag{2.67}$$

$$\frac{\partial H}{\partial \theta} = -\frac{2}{mb^2}\frac{\cos\theta}{\sin^3\theta}p_\phi^2 - |q|E_0 b\sin\theta \tag{2.68}$$

$$\frac{\partial H}{\partial \phi} = 0 \tag{2.69}$$

and the Liouville equation for the dipole is

$$\frac{\partial \rho_s}{\partial t} = -\frac{1}{2m}\left[p_x\frac{\partial \rho_s}{\partial x} + p_y\frac{\partial \rho_s}{\partial y} + p_z\frac{\partial \rho_s}{\partial z}\right] - \frac{2}{mb^2}\left[p_\theta\frac{\partial \rho_s}{\partial \theta} + \frac{1}{\sin^2\theta}p_\theta\frac{\partial \rho_s}{\partial \phi}\right] - \left[\frac{2}{mb^2}\frac{\cos\theta}{\sin^3\theta}p_\phi^2 + |q|E_0 b\sin\theta\right]\frac{\partial \rho_s}{\partial p_\theta} \tag{2.70}$$

where $\rho_s = \rho_s(x, y, z, \theta, \phi, p_x, p_y, p_z, p_\theta, p_\phi, t)$. Note that in this case $s = 10$. Equation (2.70) can be solved along the characteristics as shown in the previous example. The characteristics represent the trajectory of the dipole, i.e., displacement of the center of mass (due to some initial center of mass velocity) and rotation of the dipole about its center of mass (due to the electric field and any initial rotational motion).

2.4 Summary

The Liouville equation is called the basic equation of statistical mechanics. It merges classical mechanics with probability theory, and it describes the space-time evolution of the probability of finding an N-molecule system with a particular set of positions and velocities or, more generally, generalized coordinates and conjugate momenta. The probability function is analogous to an ordinary mass density function and the Liouville equation can be obtained in a manner analogous to the derivation of the ordinary mass conservation equation. In the case of the Liouville equation, however, the conservation balance must be done in a hyperspace representing all of the degrees of freedom or generalized coordinates and associated conjugate momenta of the system. Because the molecular information involves generalized coordinates or degrees of freedom and the associated conjugate momenta, it is the hamiltonian dynamics introduced in Chap. 1 that is directly incorporated into the Liouville equation. This incorporation provides the formal integration of probability theory and classical mechanics.

In subsequent chapters, we will show how both the equilibrium and nonequilibrium descriptions of matter follow from the Liouville equation. This will allow us to consider both subjects under a similar roof, i.e., the "Liouville roof," which will prove to have a number of advantages in applications and analyses. It will be shown that in applications it is not generally necessary to consider all of the N molecules or atoms of any particular system and, thus, a lower, *representative* number of molecules or atoms usually suffices. In the next chapter, we will consider so-called reduced forms of the Liouville equation based on a smaller, representative number of system molecules, obtained by integrating over the unnecessary set of molecules.

Problems

2.1 *Laser-Etching Process.* In certain photo-etching processes, a laser pointed at a target area $\mathbf{r}_0$ causes an emission of surface atoms with a gaussian distribution in velocities, i.e., we have the initial state for the single atom distribution function as (see Fig. 2.4)

$$\rho_1 = M\delta(\mathbf{r}_1 - \mathbf{r}_1')\frac{1}{2^{1/2}(\pi mkT)^{3/2}}\exp\left\{\sum_{i=x,y,z}\frac{p_{1i}^2}{2mkT}\right\} \quad \text{at} \quad t=0$$

For this system, obtain the solution to the Liouville equation for the time dependent, single atom distribution function under a constant external force. Show that the initial distribution function satisfies the normalization condition

$$\int\int_{\text{all } \mathbf{r}_1, \mathbf{p}_1} \rho_1 d\mathbf{r}_1 d\mathbf{p}_1 = M$$

2.2 *The Ion Cyclotron.* Consider again the solution to the single-particle Liouville equation for a single atom of charge q in a uniform magnetic field, $\mathbf{B} = B_0 e_3$ (see Fig. 2.5). The velocity-dependent Lorenz force is known to be

$$\mathbf{F}_1 = \frac{q}{m}\mathbf{p}_1 \times \mathbf{B}$$

Show that the Liouville equation for this system is given by

$$\frac{\partial \rho_1}{\partial t} + \frac{\mathbf{p}_1}{m}\cdot\frac{\partial \rho_1}{\partial \mathbf{r}_1} = \frac{q}{m}p_{1x}B_0\frac{\partial \rho_1}{\partial p_{1y}} - \frac{q}{m}p_{1y}B_0\frac{\partial \rho_1}{\partial p_{1x}}$$

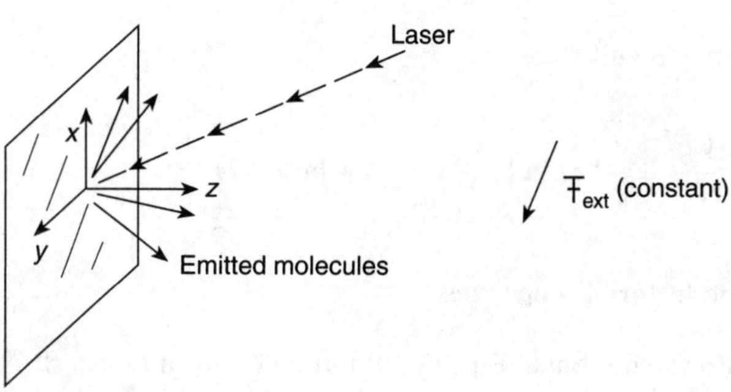

Figure 2.4 Laser-etching process.

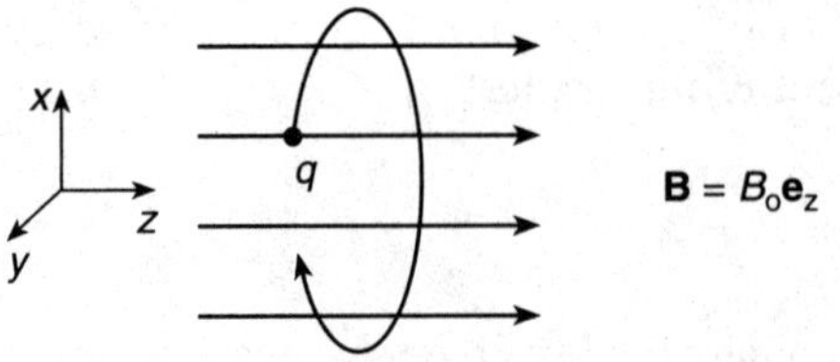

Figure 2.5 The ion cyclotron.

2.3 *General Solution to the Ion Cyclotron.* Show that the solution characteristics to Prob. 2.2 are given by

$$p_{1x} = \sqrt{C_4}\sin\left[\frac{q}{m}B_0(C_5 - t)\right]$$

$$p_{1y} = \sqrt{C_4}\cos\left[\frac{q}{m}B_0(C_5 - t)\right]$$

$$\left[p_{1x}^2 + p_{1y}^2\right]^{1/2} = \sqrt{C_4}$$

$$x_1 = \frac{\sqrt{C_4}}{m}\cos\left[\frac{q}{m}B_0(C_5 - t)\right] + C_2$$

$$y_1 = \frac{\sqrt{C_4}}{m}\sin\left[\frac{q}{m}B_0(C_5 - t)\right] + C_1$$

$$z_1 = \frac{p_{1z}t}{m} + C_3$$

where C_1, C_2, C_3, C_4, and C_5 are integration constants. For specified initial conditions, $p_{1x_0}, p_{1y_0}, p_{1z_0}, x_{1_0}, y_{1_0}, z_{1_0}$, plot $x_1(t)$ and $y_1(t)$ predicted by the characteristic equations, which should be noted, are solutions to the equations of motion for the ion.

2.4 Prove the unintuitive result that

$$\frac{\partial V_{hs}}{\partial N} \cong -V_{hs}\ln\left(\frac{N}{2}\right) \qquad \text{for large } N$$

(*Hint*: See Ref. 1 on factorial properties.)

2.5 Substitute Hamilton's equations, Eqs. (1.104) and (1.105) into Eq. (2.7) and simplify to obtain Eq. (2.8).

References

1. M. Abramowitz and I. A. Stegun, *Handbook of Mathematical Functions*, Dover Publications, New York, 1970.
2. G. F. Carrier and C. E. Pearson, *Partial Differential Equations. Theory and Technique*, 2nd ed., Academic Press, New York, 1988.

Further Reading

J. O. Hirschfelder, C. F. Curtiss, and R. B. Bird, *Molecular Theory of Gases and Liquids*, Wiley, New York, 1964. [Chapter 1, Sec. 4, treats the Liouville equation.]

C. Kittel, *Elementary Statistical Physics*, Wiley, New York, 1958. [An excellent introductory discussion of the Liouville equation can be found in Chap. 3.]

D. A. McQuarrie, *Statistical Mechanics*, Harper & Row, New York, 1976. [Chapter 7 treats the Liouville equation.]

Chapter

3

Reduced Density Functions and the Reduced Liouville Equation

"But neither Bartholomew Cubbins
nor King Derwin himself
nor anyone else in the Kingdom of Didd
could ever explain how the strange thing
happened.
They could say it just happened to happen
and was not very likely to happen again"

—THEODOR SEUSS GEISEL,
The 500 Hats of Bartholomew Cubbins

3.1 Overview: The Practical Importance of the Reduced Liouville Equation

As will be shown in later chapters, the equilibrium thermodynamic and nonequilibrium transport properties of gases and liquids can normally be expressed in terms of lower-ordered or reduced density functions and, in general, it is not necessary to determine the entire $6N$-dimensional density function ρ_N introduced in the last chapter. Physically, this is possible because of the typical *short-range* nature of the intermolecular interaction forces. In essence, the behavior of any given molecule in a substance is dictated by its few neighboring or surrounding molecules and not by the molecules outside its small region of force influence. It practically suffices, therefore, to examine the probability of configurations and momenta of a small group of molecules. It will be shown that the absolute lowest possible form of the reduced Liouville equation for small groups of interacting molecules leads to the famous Boltzmann transport equation, and subsequently, to Boltzmann's law of entropy increase. More specifically, Boltzmann's famous analysis is based on consideration of just two interacting molecules, and so it is most applicable to dilute gas systems described in the first chapter. As will be discussed in later chapters, the law of entropy increase, established from molecular arguments by Boltzmann, tells us a specific direction of change for a given system. The concept that systems tend to change in a particular direction has enormous consequences for both mechanical and chemical systems. For example, we know from elementary thermodynamic courses that the law of entropy increase, or the so-called second law of thermodynamics, sets important limits on the work that can be obtained as a result of energy changes of a substance. It is also interesting to note that Boltzmann's entropy arguments are based on molecular probabilities and that it is theoretically possible, according to Boltzmann, to violate the second law, but the likelihood of that happening is astronomically small—something like one over the total number of seconds the universe has existed! Developing a complete molecular description of entropy for all types of substances still remains a rather elusive but practically important and theoretically compelling aspect of statistical mechanics.

3.2 Reduced Density Functions

To begin with, we introduce a normalized density function or probability density function

$$P_N(\mathbf{r}^N, \mathbf{p}^N, t) = \frac{1}{M}\rho_N(\mathbf{r}^N, \mathbf{p}^N, t) \tag{3.1}$$

where M is the total number of members of the ensemble. Thus, the function P_N conforms to the usual requirement of probability density functions, namely[†]

$$\int\int P_N d\mathbf{r}^N d\mathbf{p}^N = 1 \tag{3.2}$$

Now consider the probability of finding only a certain small group of molecules, such as molecules (2), (4), (5), and (7), in their respective region of phase space. In particular,

$$P_h(\mathbf{r}^h, \mathbf{p}^h, t) = \int\int P_N(\mathbf{r}^N, \mathbf{p}^N, t)\, d\mathbf{r}^{N-h} d\mathbf{p}^{N-h} \tag{3.3}$$

[†]Throughout this text we use a single integral sign $\int d\mathbf{r}^N$ to denote the actual $3N$ multidimensional integral $\int \cdots \int d\mathbf{r}_1 \cdots d\mathbf{r}_N$.

where $\mathbf{r}^h = (\mathbf{r}_2, \mathbf{r}_4, \mathbf{r}_5, \mathbf{r}_7)$ and $\mathbf{p}^h = (\mathbf{p}_2, \mathbf{p}_4, \mathbf{p}_5, \mathbf{p}_7)$ in our example. Also, $d\mathbf{r}^{N-h}$ denotes all phase space except that associated with the set of molecules $\{h\}$; in our example, $d\mathbf{r}^{N-h}d\mathbf{p}^{N-h} = d\mathbf{r}_1 d\mathbf{r}_3 d\mathbf{r}_6 d\mathbf{r}_8 \ldots d\mathbf{r}_N d\mathbf{p}_1 d\mathbf{p}_3 d\mathbf{p}_6 d\mathbf{p}_8 \ldots d\mathbf{p}_N$.

Note that the probability of finding our N molecules in any order is just $N!$ times as large, or

$$f_N(\mathbf{r}^N, \mathbf{p}^N, t) \equiv N! P_N(\mathbf{r}^N, \mathbf{p}^N, t) \tag{3.4}$$

since there are N choices for the phase space location of the first molecule, $N - 1$ for the second, and so on. The function f_N applies to molecules that are indistinguishable from one another (see Fig. 3.1). Note that f_N is not normalized, i.e.,

$$\int f_N(\mathbf{r}^N, \mathbf{p}^N, t)\, d\mathbf{r}^N d\mathbf{p}^N = N! \tag{3.5}$$

Similarly, for the lower-order function $P_h(\mathbf{r}^h, \mathbf{p}^h, t)$, when order doesn't matter, it becomes

$$f_h(\mathbf{r}^h, \mathbf{p}^h, t) = \frac{N!}{(N-h)!} P_h(\mathbf{r}^h, \mathbf{p}^h, t) \tag{3.6}$$

where $[N!/(N-h)!]$ is the number of ways h objects can be chosen from N objects.[†] From Eqs. (3.5) and (3.6) it also follows that (see Prob. 3.1)

$$f_h(\mathbf{r}^h, \mathbf{p}^h, t) = [(N-h)!]^{-1} \int\int f_N(\mathbf{r}^N, \mathbf{p}^N, t)\, d\mathbf{r}^{N-h} d\mathbf{p}^{N-h} \tag{3.7}$$

In many problems we will only be interested in the configurational part of the density functions, and thus, we define a *configurational* density function as

$$P_N^c(\mathbf{r}^N, t) = \int P_N(\mathbf{r}^N, \mathbf{p}^N, t)\, d\mathbf{p}_N \tag{3.8}$$

Similarly, we have the reduced density functions in configurational space

$$P_h^c(\mathbf{r}^h, t) = \int P_N^c(\mathbf{r}^N, t)\, d\mathbf{r}^{N-h} \tag{3.9}$$

[†] N choices for the first molecule, $N - 1$ for the second, and $N - (h - 1)$ for the hth molecule. Thus, $(N)(N-1)\ldots[N-(h-1)] = N!/(N-h)!$

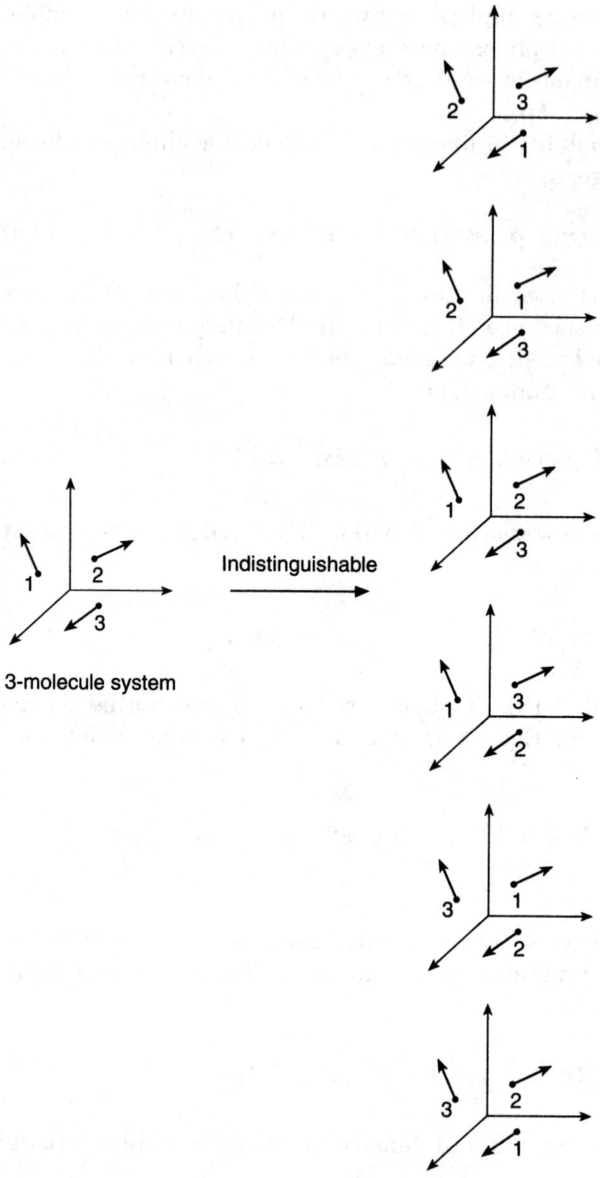

Figure 3.1 Example of the number of ways of representing a system of three indistinguishable molecules. The six indistinguishable systems are physically identical.

and, when order doesn't matter,

$$n_N(\mathbf{r}^N, t) = N! P_N^c(\mathbf{r}^N, t) \tag{3.10}$$

$$\begin{aligned} n_h(\mathbf{r}^h, t) &= \frac{N!}{(N-h)!} P_h^c(\mathbf{r}^h, t) \\ &= [(N-h)!]^{-1} \int n_N(\mathbf{r}^N, t)\, d\mathbf{r}^{N-h} \end{aligned} \tag{3.11}$$

Also note that this last expression can be written as (see Prob. 3.2)

$$n_h(\mathbf{r}^h, t) = \int f_h(\mathbf{r}^h, \mathbf{p}^h, t)\, d\mathbf{p}^h \tag{3.12}$$

In particular, $n_1(\mathbf{r}, t)\, d\mathbf{r}_1$ is, aside from a constant, the probability of finding *any one molecule* in the space between $\mathbf{r}_1$ and $\mathbf{r}_1 + d\mathbf{r}_1$. This is the probability interpretation of the ordinary macroscopic number density $n(\mathbf{r}, t)$ found in the equation of continuity to be given in Chap. 5.

Example 3.1 Lottery Chances As a simple example of the practical use of combinatorial algebra, consider the odds of selecting six numbers, in any order, from a set of 49 numbers. In general, we have 49 choices for the first number, 48 for the second number, and so on leading to $49 \times 48 \times 47 \times 46 \times 45 \times 44$ ($\cong 1.0068 \times 10^{10}$) possible combinations. However, the order of our six numbers does not matter; so, we have a factor of 6! too many combinations (six choices for the first, five for the second, and so on). Thus, the total actual number of combinations is, approximately

$$\frac{1.0068 \times 10^{10}}{6!} = 13,984,816$$

making our odds of winning the lottery with a single ticket about one in fourteen million!

3.3 The Reduced Liouville Equation

Consider a reduced form of the Liouville equation for a set of molecules $\{s\} = \{1, 2, 3, \ldots, s\}$ that can be obtained by integrating the Liouville equation over the phase space of the other $\{N - s\}$ set of molecules. Referring to Eq. (2.6) written in terms of f_N, term by term, we obtain for the first term

$$\begin{aligned} \int\!\!\int \frac{\partial f_N}{\partial t} d\mathbf{r}^{N-s} d\mathbf{p}^{N-s} &= \frac{\partial}{\partial t} \int\!\!\int f_N d\mathbf{r}^{N-s} d\mathbf{p}^{N-s} \\ &= (N-s)!\, \frac{\partial f_s}{\partial t} \end{aligned} \tag{3.13}$$

where we have used Eq. (3.6). For the second term,

$$\sum_{i=1}^{N}\int\int\left(\frac{\mathbf{p}_i}{m_i}\cdot\frac{\partial f_N}{\partial \mathbf{r}_i}\right)d\mathbf{r}^{N-s}d\mathbf{p}^{N-s}=(N-s)!\sum_{i=1}^{s}\left(\frac{\mathbf{p}_i}{m_i}\cdot\frac{\partial f_s}{\partial \mathbf{r}_i}\right)$$
$$+\sum_{i=s+1}^{N}\frac{1}{m_i}\int\int\frac{\partial}{\partial \mathbf{r}_i}\cdot[\mathbf{p}_i f_N]\,d\mathbf{r}^{N-s}d\mathbf{p}^{N-s}=(N-s)!\sum_{i=1}^{s}\left(\frac{\mathbf{p}_i}{m_i}\cdot\frac{\partial f_s}{\partial \mathbf{r}_i}\right)$$
$$+\sum_{i=s+1}^{N}\frac{1}{m_i}\int\int\left\{\int\int\frac{\partial}{\partial \mathbf{r}_i}\cdot[\mathbf{p}_i f_N]d\mathbf{r}_i d\mathbf{p}_i\right\}d\mathbf{r}^{N-s-i}\,d\mathbf{p}^{N-s-i}$$
$$(3.14)$$

The positional integral part of the term in curly brackets can be rewritten using the divergence or Gauss' theorem, written generically as

$$\int_{V_e}\left[\frac{\partial}{\partial \mathbf{e}}\cdot(a\mathbf{b})\right]d\,\mathbf{e}=\int_{S_e}a\mathbf{b}\cdot\mathbf{n}_e\,d\,S \qquad (3.15)$$

where $\mathbf{n}_e$ is a unit normal vector directed outward from the surface S_e enclosing the volume V_e, leading to

$$\int\frac{\partial}{\partial \mathbf{r}_i}\cdot[\mathbf{p}_i f_N]d\mathbf{r}_i=\lim_{S_{r_i}\to\infty}\left[\int_{S_{r_i}}f_N\mathbf{p}_i\cdot\mathbf{n}_{r_i}\,d\,S_{r_i}\right] \qquad (3.16)$$

where the integration is performed over the surface enclosing the entire $\mathbf{r}_i$ space. Now, it is assumed that the N-particle density function has the following properties:

$$\left.\begin{aligned}f_N(\mathbf{r}^N,\mathbf{p}^N,t)&\to 0\\ \mathbf{p}_i f_N(\mathbf{r}^N,\mathbf{p}^N,t)&\to 0\\ \mathbf{F}_i f_N(\mathbf{r}^N,\mathbf{p}^N,t)&\to 0\end{aligned}\right\}\ \text{as } \mathbf{r}^N,\mathbf{p}^N\to\infty \qquad (3.17)$$

This is the usual behavior of probability density functions, i.e., asymptotically the probability of finding molecules at distances or momenta approaching infinity should decay rapidly to zero.

With the properties, Eq. (3.17), the integral term, Eq. (3.16), is identically zero and the second term on the right-hand side of Eq. (3.14) vanishes. Thus, we are left with

$$\sum_{i=1}^{N}\int\int\left(\frac{\mathbf{p}_i}{m_i}\cdot\frac{\partial f_N}{\partial \mathbf{r}_i}\right)d\mathbf{r}^{N-s}d\mathbf{p}^{N-s}=(N-s)!\sum_{i=1}^{s}\left(\frac{\mathbf{p}_i}{m_i}\cdot\frac{\partial f_s}{\partial \mathbf{r}_i}\right) \qquad (3.18)$$

The remaining term in the Liouville equation is

$$\sum_{i=1}^{N} \int \int \left(\mathbf{F}_i \cdot \frac{\partial f_N}{\partial \mathbf{p}_i} \right) d\mathbf{r}^{N-s} d\mathbf{p}^{N-s}$$

$$= \sum_{i=1}^{s} \frac{\partial}{\partial \mathbf{p}_i} \cdot \int \mathbf{F}_i f_N d\mathbf{r}^{N-s} d\mathbf{p}^{N-s} + \sum_{i=s+1}^{N} \int \frac{\partial}{\partial \mathbf{p}_i} \cdot [\mathbf{F}_i f_N] d\mathbf{r}^{N-s} d\mathbf{p}^{N-s} \tag{3.19}$$

Again, using Gauss' theorem, Eq. (3.15), and the properties in Eq. (3.17), the second term on the right-hand side of Eq. (3.19) is zero. Note the assumption that $\mathbf{F}_i = \mathbf{F}_i(\mathbf{r}^N)$, i.e., the force on any molecule is assumed to depend only on the position vectors of all molecules.

Substituting Eqs. (3.13), (3.18), and (3.19) into Eq. (2.6) integrated over $d\mathbf{r}^{N-s} d\mathbf{p}^{N-s}$ space, leads to the *reduced* Liouville equation

$$\frac{\partial f_s}{\partial t} + \sum_{i=1}^{s} \left(\frac{\mathbf{p}_i}{m_i} \cdot \frac{\partial f_s}{\partial \mathbf{r}_i} \right) + \frac{1}{(N-s)!} \sum_{i=1}^{s} \frac{\partial}{\partial \mathbf{p}_i} \cdot \int \int \mathbf{F}_i f_N d\mathbf{r}^{N-s} d\mathbf{p}^{N-s} = 0 \tag{3.20}$$

The force acting on the ith molecule can be written as the gradient of a potential as

$$\mathbf{F}_i = -\frac{\partial}{\partial \mathbf{r}_i} \Phi^N(\mathbf{r}^N) \tag{3.21}$$

Furthermore, the potential can often be approximated by a sum over the pair interaction potentials as

$$\mathbf{F}_i = -\frac{\partial}{\partial \mathbf{r}_i} \Phi^N(\mathbf{r}^N) \approx -\sum_{\substack{j=1 \\ j \neq i}}^{N} \frac{\partial}{\partial \mathbf{r}_i} \phi(\mathbf{r}_{ij}) \tag{3.22}$$

where the pair potential $\phi(\mathbf{r}_{ij})$ is the interaction potential introduced in Chap. 1, between any two molecules in the system; i.e., the effects of three or more body interactions on the pair potential expression are neglected.[†] Equation (3.22) constitutes the so-called pairwise additivity approximation. The specific forms of $\phi(r_{ij})$, such as the Lennard-Jones (L-J) and coulombic potentials, were discussed in Chap. 1. Pairwise

[†]Note that we have changed notation from $u(r)$ used in Chap. 1 to $\phi(r)$ used here and in all remaining chapters to avoid any confusion with the macroscopic internal energy (U) to be introduced in the next chapter.

additivity is exact for coulombic interactions, but approximate for L-J potentials.†

Substituting Eq. (3.22) into Eq. (3.20) gives

$$\frac{\partial f_s}{\partial t} + \sum_{i=1}^{s}\left[\left(\frac{\mathbf{p}_i}{m_i}\cdot\frac{\partial f_s}{\partial \mathbf{r}_i}\right) - \sum_{\substack{j=1\\ j\neq i}}^{s}\left(\frac{\partial \phi(\mathbf{r}_{ij})}{\partial \mathbf{r}_i}\cdot\frac{\partial f_s}{\partial \mathbf{p}_i}\right)\right]$$
$$= \frac{1}{(N-s)!}\sum_{i=1}^{s}\frac{\partial}{\partial \mathbf{p}_i}\cdot\left[\sum_{j=s+1}^{N}\int\!\!\int\frac{\partial \phi(\mathbf{r}_{ij})}{\partial \mathbf{r}_i} f_N d\mathbf{r}^{N-s} d\mathbf{p}^{N-s}\right] \tag{3.23}$$

The integration on the right-hand side can be performed over all the $(\mathbf{r}^{N-s},\ \mathbf{p}^{N-s})$ space except the $(s+1)$ molecule. Thus, Eq. (3.23) becomes (see Prob. 3.4)

$$\frac{\partial f_s}{\partial t} + \sum_{i=1}^{s}\left[\left(\frac{\mathbf{p}_i}{m_i}\cdot\frac{\partial f_s}{\partial \mathbf{r}_i}\right) - \sum_{\substack{j=1\\ j\neq i}}^{s}\left(\frac{\partial \phi(\mathbf{r}_{ij})}{\partial \mathbf{r}_i}\cdot\frac{\partial f_s}{\partial \mathbf{p}_i}\right)\right]$$
$$= \sum_{i=1}^{s}\int\!\!\int\left(\frac{\partial \phi(\mathbf{r}_{i,s+1})}{\partial \mathbf{r}_i}\cdot\frac{\partial f_{s+1}}{\partial \mathbf{p}_i}\right) d\mathbf{r}_{s+1} d\mathbf{p}_{s+1} \tag{3.24}$$

Equation (3.24) is the reduced Liouville equation for pairwise additive interaction forces. Note that this is an integro-differential equation, where the evolution of the f_s distribution depends on the next higher-order f_{s+1} distribution. This is known as the BBGKY hierarchy (named after its originators: Bogoliubov, Born, Green, Kirkwood, and Yvon; see the Further Reading section at the end of this chapter).

Interlude 3.1 Contracted Density Functions and the Loss of Information‡ Probability density functions depend on *random* variables that are the independent variables in the system. For example, in manufacturing we could speak of the probability P of producing a part (say, a sprocket) with a certain diameter (call this variable x_1) and a certain mass (call this variable x_2); thus, we have $P = P(x_1, x_2)$. Formally, $P(x_1, x_2)dx_1dx_2$ is the probability of

†See the Further Reading section at the end of Chap. 1 for more details. In the case of van der Waals forces, which constitute the attractive part of the L-J potential, the presence of a third body slightly alters the electron distribution functions of the interacting two-body pair. Pairwise additivity usually provides a very good prediction of macroscopic behavior, even in concentrated systems, except possibly near points of phase transition or critical phenomena.

‡For a more extensive treatment of probability density functions, see Ref. 1.

finding the sprocket with a diameter and mass between x_1 and $x_1 + dx_1$ and x_2 and $x_2 + dx_2$, respectively.

Now, the probability of finding our sprocket with only a certain diameter is given by the "contracted" density function $P(x_1)$

$$P(x_1) = \int_{\text{all } x_2} P(x_1, x_2) dx_2$$

where the integration is performed over all of the x_2 space. $P(x_1)$ no longer contains any information about the mass of the sprocket x_2 and we say that the information has been *lost*.

If we look at the behavior of any one particular molecule in a system of molecules, its dynamics is only influenced by nearby molecules, or its nearest neighbors, provided that the interaction forces decay rapidly with intermolecular separation distance (the usual case). Thus, it is not always necessary to know the full N-body distribution function, and a (much) lower-order distribution function involving only the nearest neighbors will often suffice. Although we lose information by such a contraction, it is information that, for all intents and purposes, has no bearing on the problem. The analogy with our manufacturing example would be that the sprocket mass does not affect its performance and is, therefore, information that is not needed. Ultimately, it's the "physics" of the problem that dictates what information is important or not important.

3.4 The Boltzmann Transport Equation

The lowest form of Eq. (3.24) is obtained by setting $s = 1$, i.e., we have contracted the Liouville equation over all space except one molecule. Thus,

$$\frac{\partial f_1}{\partial t} + \frac{\mathbf{p}_1}{m_1} \cdot \frac{\partial f_1}{\partial \mathbf{r}_1} = \int\int \frac{\partial \phi(\mathbf{r}_1, \mathbf{r}_2)}{\partial \mathbf{r}_1} \cdot \frac{\partial f_2(\mathbf{r}_1, \mathbf{r}_2, \mathbf{p}_1, \mathbf{p}_2, t)}{\partial \mathbf{p}_1} d\mathbf{r}_2 \, d\mathbf{p}_2 \tag{3.25}$$

Now, for so-called "dilute gases" we need only consider the behavior of any two molecules, i.e., we truncate the BBGKY hierarchy at $s = 2$ and set $f_3 = 0$. Thus, we also have

$$\frac{\partial f_2}{\partial t} + \sum_{i=1}^{2} \left[\frac{\mathbf{p}_i}{m_i} \cdot \frac{\partial f_2}{\partial \mathbf{r}_i} - \sum_{\substack{j=1 \\ j \neq i}}^{2} \left(\frac{\partial \phi(\mathbf{r}_{ij})}{\partial \mathbf{r}_i} \cdot \frac{\partial f_2}{\partial \mathbf{p}_i} \right) \right] = 0 \tag{3.26}$$

Now, the following assumptions are invoked:[†]

1. The molecular pair function f_2 is only weakly dependent on time ($\partial f_2 / \partial t \cong 0$).

[†]This development is due to Ref. 2. More formal arguments will be given in Chap. 6.

2. The spatial part of the molecular pair function depends primarily on the separation distance vector $\mathbf{r}_{12} = \mathbf{r}_2 - \mathbf{r}_1$, i.e.,

$$f_2(\mathbf{r}_1, \mathbf{r}_2, \mathbf{p}_1, \mathbf{p}_2) \cong f_2(\mathbf{r}_{12}, \mathbf{p}_1, \mathbf{p}_2) \tag{3.27}$$

and, thus,

$$\frac{\partial f_2}{\partial \mathbf{r}_1} \cong -\frac{\partial f_2}{\partial \mathbf{r}_2} \tag{3.28}$$

Under these assumptions, Eq. (3.26) becomes

$$\frac{\mathbf{p}_1}{m_1} \cdot \frac{\partial f_2}{\partial \mathbf{r}_1} + \frac{\mathbf{p}_2}{m_2} \cdot \frac{\partial f_2}{\partial \mathbf{r}_2} + \mathbf{F}_{12} \cdot \frac{\partial f_2}{\partial \mathbf{p}_1} + \mathbf{F}_{21} \cdot \frac{\partial f_2}{\partial \mathbf{p}_2} = 0 \tag{3.29}$$

where we have used $\mathbf{F}_{12} = -\partial\phi(\mathbf{r}_{12})/\partial\mathbf{r}_1$ and $\mathbf{F}_{21} = -\partial\phi(\mathbf{r}_{12})/\partial\mathbf{r}_2$.

Following the methods of Chap. 2, it can be readily shown that the characteristic equations of Eq. (3.29) lead to the following relationships (Prob. 3.8)

$$\mathbf{p}_1 + \mathbf{p}_2 = C_1 \tag{3.30}$$

$$\frac{1}{4m}(\mathbf{p}_{21} \cdot \mathbf{p}_{21}) + \phi(|\mathbf{r}_{12}|) = C_2 \tag{3.31}$$

where $\mathbf{p}_{21} = \mathbf{p}_1 - \mathbf{p}_2$. Combining these last two equations gives

$$\frac{1}{2m_1}(\mathbf{p}_1 \cdot \mathbf{p}_1) + \frac{1}{2m_2}(\mathbf{p}_2 \cdot \mathbf{p}_2) + \phi(|\mathbf{r}_{12}|) = C_3 \tag{3.32}$$

where C_1, C_3, and C_3 are constants. Equations (3.30) and (3.32) are the statements of conservation of linear momentum and conservation of total energy of the two-body system, respectively.

First, rearranging Eq. (3.29) using Eq. (3.25), then substituting into the integral term gives

$$\int\int \left[\frac{\mathbf{p}_1}{m_1} \cdot \frac{\partial f_2}{\partial \mathbf{r}_1} + \frac{\mathbf{p}_2}{m_2} \cdot \frac{\partial f_2}{\partial \mathbf{r}_2} + \mathbf{F}_{21} \cdot \frac{\partial f_2}{\partial \mathbf{p}_2}\right] d\mathbf{r}_2\, d\mathbf{p}_2$$

where the third term of the integrand involving the interaction force can be shown to vanish by application of Gauss' theorem and the properties, Eq. (3.17). The remaining two terms can be combined using Eq. (3.28), leading to

$$\int\int \left(\frac{\mathbf{p}_2}{m_2} - \frac{\mathbf{p}_1}{m_1}\right) \cdot \frac{\partial f_2}{\partial \mathbf{r}_2}\, d\mathbf{r}_2\, d\mathbf{p}_2$$

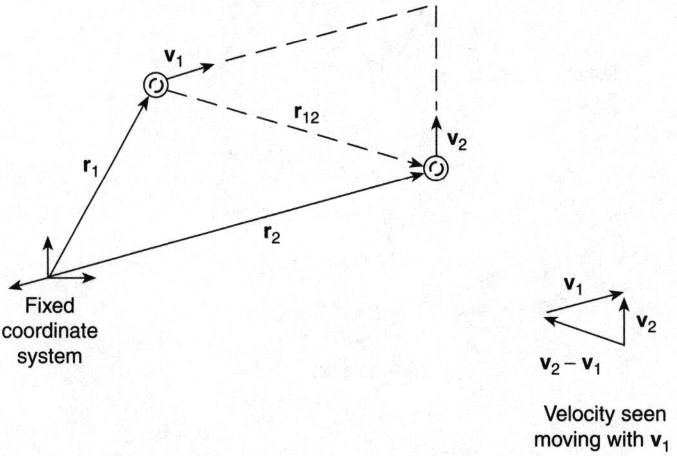

Figure 3.2 Illustration of the relative velocity and relative positional vectors for two interacting particles.

Transforming to a relative set of coordinates $\mathbf{r}_{12} = \mathbf{r}_2 - \mathbf{r}_1$ and $\mathbf{p}_{12} = \mathbf{p}_2 - \mathbf{p}_1(\mathbf{v}_{12} = \mathbf{v}_2 - \mathbf{v}_1)$ the integral can be written as (see Fig. 3.2)

$$\int\int \mathbf{v}_{12} \cdot \frac{\partial f_2}{\partial \mathbf{r}_{12}} d\mathbf{r}_{12}\, d\mathbf{p}_{12}$$

Using Gauss' theorem

$$\int\int \mathbf{v}_{12} \cdot \frac{\partial f_2}{\partial \mathbf{r}_{12}} d\mathbf{r}_{12}\, d\mathbf{p}_{12} = \int\int_{s_{12}} \mathbf{v}_{12} \cdot f_2\, d\mathbf{S}_{12}\, d\mathbf{p}_{12} \tag{3.33}$$

Now consider a coordinate system with molecule (1) at the origin (Fig. 3.2). The surface S_{12} is taken to be just outside the range of the interaction force $\mathbf{F}_{21}$, as shown in Fig. 3.3. For an incoming molecule (2) $(d\mathbf{S}_{12} \cdot \mathbf{v}_{12} < 0)$ we have no correlations between molecule (1) and molecule (2) along the surface and, thus,

$$f_2 = f_1(\mathbf{r}_1, \mathbf{p}_1, t) f_1(\mathbf{r}_2, \mathbf{p}_2, t) \qquad \mathbf{v}_{12} \cdot d\mathbf{S}_{12} < 0 \tag{3.34}$$

Similarly, for outgoing molecules along the surface

$$f_2 = f_1(\mathbf{r}_1', \mathbf{p}_1', t) f_1(\mathbf{r}_2', \mathbf{p}_2', t) \qquad \mathbf{v}_{12} \cdot d\mathbf{S}_{12} > 0 \tag{3.35}$$

Finally, we take the surface of integration (outside the interaction force range) to be a cylinder, as shown in Fig. 3.4. The axis of the cylinder is taken to be parallel to the relative velocity vector $\mathbf{v}_{12}$. This is the velocity of molecule (2) observed moving along with molecule (1). The cylinder radial coordinate b is called the *impact parameter*. It is

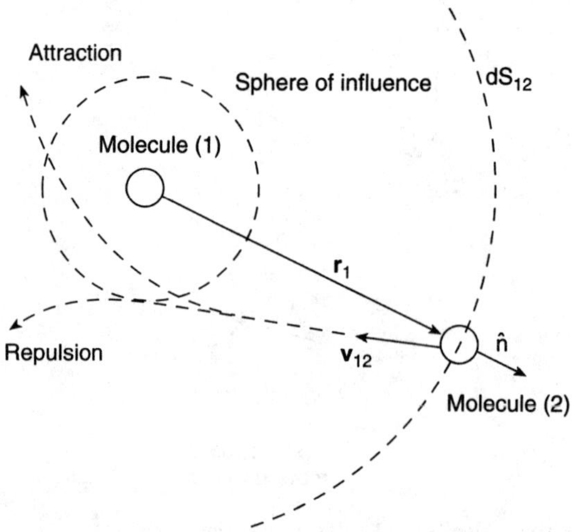

Figure 3.3 Interaction of two molecules showing the "sphere of influence" of molecule (1). Inside the sphere of influence, molecule (2) moves closer to molecule (1) under an attractive force and away from molecule (1) under a repulsive force.

the distance of closest approach of molecule (2) to molecule (1) in the absence of an interaction force. We denote the outer radius of the cylinder as b_0. This is the distance at which the interaction force becomes negligible. Along the outer shell of the cylinder, $\mathbf{v}_{12} \cdot d\mathbf{S}_{12} = 0$. Thus, for a selected value of $\mathbf{p}_{12}$, the only contribution to the surface integral

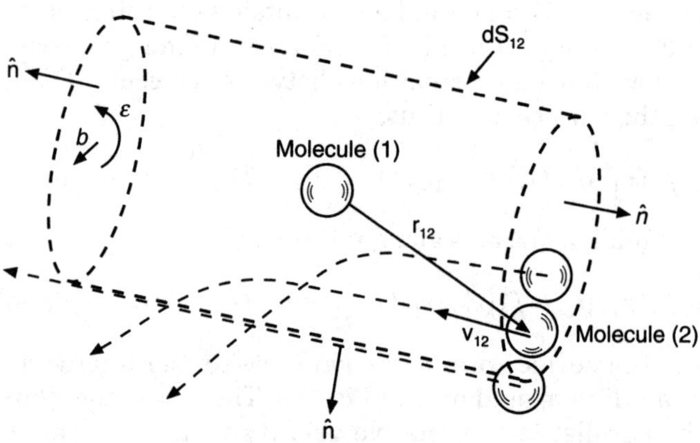

Figure 3.4 Illustration and terminology of the cylindrical control volume used for the purpose of integration.

comes from the ends of the cylinder, i.e.,

$$\int\int_{S_{12}} \mathbf{v}_{12} \cdot f_2 \, d\mathbf{S}_{12} \, d\mathbf{p}_{12}$$
$$= 2\pi \int_{-\infty}^{\infty} \int_{0}^{b_0} v_{12}[f_1(\mathbf{r}_1', \mathbf{p}_1', t) f_1(\mathbf{r}_2', \mathbf{p}_2', t) - f_1(\mathbf{r}_1, \mathbf{p}_1, t) f_1(\mathbf{r}_2, \mathbf{p}_2, t)] \times b \, db \, d\mathbf{p}_{12} \tag{3.36}$$

where v_{12} is the relative molecular speed ($|\mathbf{v}_{12}|$) and we have used Eqs. (3.34) and (3.35). Note that the cylinder length must be large enough so that the interaction force is also zero at the ends. Equation (3.36) is called the *Boltzmann collision integral*.

Putting Eq. (3.36) into the right-hand side of Eq. (3.25) gives the famous *Boltzmann transport equation*.†

$$\frac{\partial f_1}{\partial t} + \frac{\mathbf{p}_1}{m_1} \cdot \frac{\partial f_1}{\partial \mathbf{r}_1}$$
$$= 2\pi \int_{-\infty}^{\infty} \int_{0}^{b_0} v_{12}[f_1(\mathbf{r}_1', \mathbf{p}_1', t) f_1(\mathbf{r}_2', \mathbf{p}_2', t) - f_1(\mathbf{r}_1, \mathbf{p}_1, t) f_1(\mathbf{r}_2, \mathbf{p}_2, t)] \times b \, db \, d\mathbf{p}_{12} \tag{3.37}$$

Solution to the Boltzmann transport equation enables the determination of the transport properties (such as viscosity, thermal conductivity, and diffusion) of dilute gases, as will be shown in Chap. 6. (Also, see the Further Reading section at the end of this chapter.)

Finally, we note an important property of the conservation equations, Eqs. (3.30) and (3.32), in the context of the Boltzmann collision integral. Letting primes denote postcollisional values and unprimed denote the precollisional values, as in the Boltzmann collision integral, the conservation equations, when applied to a collision, lead to

$$\mathbf{p}_1 + \mathbf{p}_2 = \mathbf{p}_1' + \mathbf{p}_2' \tag{3.38}$$

$$\frac{1}{2m_1} p_1^2 + \frac{1}{2m_2} p_2^2 = \frac{1}{2m_1} p_1'^2 + \frac{1}{2m_2} p_2'^2 \tag{3.39}$$

The latter equation follows from Eq. (3.32) since the interaction potential is zero outside the cylinder of radius b_0. In general, quantities

†A more formal derivation of the Boltzmann equation using the method of multiple time scales is given in Ref. 3. More analysis and discussion will be given in Chap. 6.

$\psi_i (i = 1, 2)$ that satisfy the condition

$$\psi_1 + \psi_2 - \psi_1' - \psi_2' = 0 \tag{3.40}$$

are called *collisional invariants*. We have shown that the mass, momentum, and energy are the collisional invariants in a two-body encounter.

It suffices to say that extensions of the reduced Liouville equation to higher orders, necessary for dense gases and liquids, become extremely cumbersome. Approximate solutions to the reduced Liouville equation for nonequilibrium dense gas and liquid systems will be considered in more detail in Chap. 6. In the next chapter, however, we will show that exact analytical solutions to the Liouville equation, and its reduced forms, are indeed possible for systems at equilibrium.

Before leaving this chapter, we briefly look at an important quantity known as Boltzmann's entropy, and we will examine reduced forms of the Liouville equation in generalized coordinates.

3.5 Boltzmann's Entropy

Perhaps one of the most important consequences of the Boltzmann equation, discovered by Boltzmann himself, was that it led to a kinetic molecular theory basis for the law of entropy increase. This law is contained within the so-called Boltzmann H-theorem. Here we will work directly with the Boltzmann entropy S_B rather than the H-function used in Boltzmann's original work.[†] We define *a local* Boltzmann entropy $S_B(\mathbf{r}_1, t)$ as

$$S_B(\mathbf{r}_1, t) \equiv -\frac{k}{m} \int f_1(\mathbf{r}_1, \mathbf{p}_1, t) \ln f_1(\mathbf{r}_1, \mathbf{p}_1, t)\, d\mathbf{p}_1 \tag{3.41}$$

Now, the total change in S_B with time is given by[‡]

$$\frac{dS_B}{dt} = -\frac{k}{m} \int (1 + \ln f_1) \frac{df_1}{dt}\, d\mathbf{p}_1 = -\frac{k}{m} \int \ln f_1 \frac{df_1}{dt}\, d\mathbf{p}_1 \tag{3.42}$$

Using the chain rule for df_1/dt and then the Boltzmann equation, Eq. (3.37), we have

$$\begin{aligned} \frac{df_1}{dt} &= \frac{\partial f_1}{\partial t} + \frac{\mathbf{p}_1}{m_1} \cdot \frac{\partial f_1}{\partial \mathbf{r}_1} + \mathbf{F}_1 \cdot \frac{\partial f_1}{\partial \mathbf{p}_1} \\ &= 2\pi \int \int [f_1(1') f_1(2') - f_1(1) f_1(2)] v_{12}\, b\, db\, d\mathbf{p}_{12} \end{aligned} \tag{3.43}$$

[†]Many of Boltzmann's contributions can be found in Ref. 4.

[‡]$\int \frac{df_1}{dt} d\mathbf{p}_1 = \frac{d}{dt} \int f_1 d\mathbf{p}_1 = \frac{dn_1}{dt} = 0$, by the equation of continuity to be shown in Chap. 5.

where we have simplified the notation as $f_1(\mathbf{r}_1', \mathbf{p}_1', t) \equiv f_1(1')$, and the like. Substituting Eq. (3.43) into Eq. (3.41) gives

$$\frac{dS_B}{dt} = -\frac{k}{m}2\pi \int\int [\ln f_1(1)][f_1(1')f_1(2') - f_1(1)f_1(2)]v_{12}bdbd\,\mathbf{p}_{12}d\,\mathbf{p}_1 \tag{3.44}$$

Now, for identical molecules (1) and (2), Eqs. (3.42) and (3.43) are invariant under a change of indices 1 → 2, i.e., we also have

$$\frac{dS_B}{dt} = -\frac{k}{m}2\pi \int\int\int [\ln f_1(2)][f_1(2')f_1(1') - f_1(2)f_1(1)]v_{21}\,bdbd\,\mathbf{p}_{21}d\,\mathbf{p}_2 \tag{3.45}$$

Noting that $v_{12} = v_{21}$ and $d\,\mathbf{p}_{21}d\,\mathbf{p}_2 = d\,\mathbf{p}_{12}d\,\mathbf{p}_1$,† we can add Eqs. (3.44) and (3.45) to obtain

$$\frac{dS_B}{dt} = -\frac{k}{m}\frac{1}{2}(2\pi)\int\int\int \ln[f_1(1)f_1(2)][f_1(1')f_1(2') - f_1(1)f_1(2)]v_{12}\,bdbd\,\mathbf{p}_{12}d\,\mathbf{p}_1 \tag{3.46}$$

Now consider the interchange of primed and unprimed variables in Eq. (3.46). Our cylinder for integration must now be aligned along the $\mathbf{v}_{12}'$ direction as shown in Fig. (3.5). The entropy change for the system will be the same whether we integrate the collision over primed or unprimed quantities. Also, the relative speed v_{12} will not change either [see Eq. (3.31)]. Thus, from Eq. (3.46) we must have

$$\frac{dS_B}{dt} = -\frac{k}{m}\frac{1}{2}\int\int\int \ln[f_1(1')f_1(2')][f_1(1)f_1(2) - f_1(1')f_1(2')]v_{12}\,bd\,bd\,\mathbf{p}_{12}'d\,\mathbf{p}_1' \tag{3.47}$$

There are a couple of different ways of showing that the integral over $d\,\mathbf{p}_{12}'d\,\mathbf{p}_1'$ is identical to the integral over $d\,\mathbf{p}_{12}d\,\mathbf{p}_1$. The simplest method is based on the *conservation of extension in phase space* already presented in Chap. 2, which states that there is no change with time in the volume of phase space occupied by the phase points.‡ Thus, we can

†The relationship $d\,\mathbf{p}_{12}d\,\mathbf{p}_1 = d\,\mathbf{p}_{21}d\,\mathbf{p}_2$, follows from the transformations $\mathbf{p}_1 = \mathbf{p}_{21}+\mathbf{p}_2$ and $\mathbf{p}_{12} = -\mathbf{p}_{21}$, where $\mathbf{p}_{12} = \mathbf{p}_2 - \mathbf{p}_1$. The absolute value of the jacobian of this transformation, $(\mathbf{p}_{12}, \mathbf{p}_1) \to (\mathbf{p}_{21}, \mathbf{p}_2)$, is unity. For a discussion of integral transformations, see Ref. 5.

‡Also, see Curtiss, Hirshfelder, and Bird in the Further Reading section at the end of this chapter, p. 448, Eq. 7.1–14 and p. 85. The second method involves a more elaborate accounting of the collisional mechanics; see Harris in the Further Reading section at the end of this chapter, p. 36.

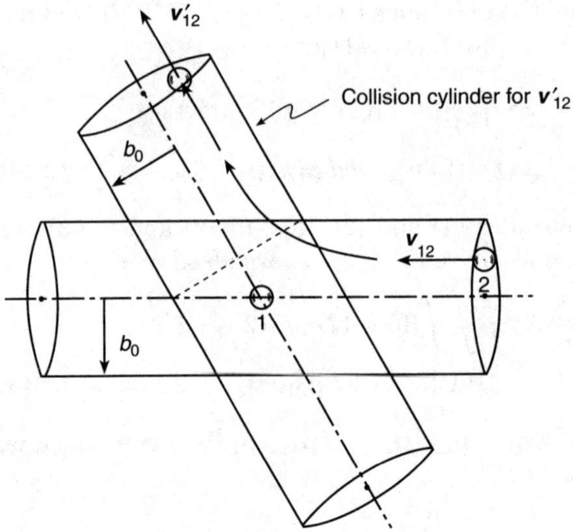

Figure 3.5 Cylinder for integrations over primed or postcollisional quantities.

combine Eqs. (3.46) and (3.47) to obtain

$$\frac{dS_B}{dt} = \frac{k}{m}\frac{1}{2}\int\int\int \{\ln[f_1(1')f_1(2')] - \ln[f_1(1)f_1(2)]\} \times [f_1(1')f_1(2') - f_1(1)f_1(2)]v_{12}\, bd\, bd\, \mathbf{p}_{12} d\,\mathbf{p}_1 \tag{3.48}$$

Now, we must have

$$(\ln x - \ln y)(x - y) \geq 0 \tag{3.49}$$

which follows because both factors always have the same sign. Thus,

$$\frac{dS_B}{dt} \geq 0 \tag{3.50}$$

which is the familiar *law of entropy increase*.

We will return to Boltzmann's entropy in Chap. 5. There is yet another molecular theory definition of entropy, which will be introduced in the next chapter known as *Gibbs' entropy*. Comparisons of Gibbs' to Boltzmann's entropy, as well as the introduction of a general form that encompasses both definitions, will be given in Chap. 5. A general entropy conservation equation, valid for both gases and liquids, will also be derived in Chap. 5, and the law of entropy increase will be rederived for gases in a more straightforward manner in Chap. 6.

Finally, we note that for a gas at equilibrium, $dS_B/dt = 0$ and, thus,

$$\ln f_1(1') + \ln f_1(2') - \ln f_1(1) - \ln f_1(2) = 0 \tag{3.51}$$

We see that, at equilibrium, $\ln f_1$ follows the form of a collisional invariant, Eq. (3.40). The most general form of $\ln f_1$ is, therefore, a linear combination of all of the collisional invariants of the two-body problem [Eqs. (3.38) and (3.39), and conservation of mass]. Thus,

$$\ln f_1(\mathbf{p}_1) = C_1 m_1 + \mathbf{C}_2 \cdot \mathbf{p}_1 + C_3 p_1^2/2m_1 \qquad \text{(equilibrium only)} \tag{3.52}$$

or

$$f_1(\mathbf{p}_1) = e^{C_1 m_1} e^{\mathbf{C}_1 \cdot \mathbf{p}_1} e^{C_3 p_1^2/2m_1} \qquad \text{(equilibrium only)} \tag{3.53}$$

The constants are determined by the conditions

$$\int f_1(\mathbf{r}_1, \mathbf{p}_1, t)\, d\mathbf{p}_1 = n_0 \qquad \text{(constant equilibrium number density)} \tag{3.54}$$

$$\int_{-\infty}^{\infty} \mathbf{p}_1 f_1(\mathbf{r}_1, \mathbf{p}_1, t)\, d\mathbf{p}_1 = 0 \qquad \text{(no net momentum)} \tag{3.55}$$

and

$$\int_{-\infty}^{\infty} p_1^2 f_1(\mathbf{r}_1, \mathbf{p}_1, t)\, d\mathbf{p}_1 = \frac{3}{2} n_0 kT \qquad \text{(definition of temperature)} \tag{3.56}$$

Equation (3.56) is the dilute gas kinetic theory definition of temperature where k is Boltzmann's constant. From these relationships, the constants in Eq. (3.53) can be uniquely determined, leading to (Prob. 3.10)

$$f_1(\mathbf{p}_1) = n_0\, (2\pi mkT)^{-3/2} \exp\left\{ -\frac{p_1^2}{2mkT} \right\} \tag{3.57}$$

This relationship should be familiar to the reader—it is the famous maxwellian velocity distribution formula for a gas at equilibrium.

Interlude 3.2 Poincaré Recurrence Times We have seen that Boltzmann's entropy theorem leads not only to an expression for the equilibrium distribution function, but also to a specific direction of change with time or *irreversibility* for a system of particles or molecules. The entropy theorem states that the entropy of a closed system can never decrease; so, whatever entropy state the system is in, it will always change to a higher entropy state. At that time, Boltzmann's entropy theorem was viewed to be contradictory to a well-known theorem in dynamics due to Poincaré. This theorem states that "in a system of material particles under the influence of forces which depend only on the spatial coordinates, a given initial state must in general recur, not exactly, but to any desired degree of accuracy infinitely often, provided the system always remains in the finite part of phase space." A straightforward derivation of Poincaré's theorem can be found in the classic paper by Chandrasekhar.† Poincaré's theorem seems to contradict Boltzmann's irreversibility result.

Boltzmann attempted to resolve this paradox by considering the system particle dynamics from a probability sense. He estimated the Poincaré recurrence time for a cubic centimeter of air containing about 10^{18} molecules. His calculated result is that the Poincaré recurrence time, i.e., the average time for the system to pass back through the initial state is on the order of the age of the universe!

As will be discussed in more detail in the next chapter, investigations in theoretical particle dynamics over the years have shed great light on so-called irreversible behavior. Several requirements on interacting particle systems are necessary for a system to achieve an equilibrium, or highly "mixed" state, and as we will see in Chap. 6, irreversibility is a direct result of the system's ability to achieve such a state. A great deal of insight into mixing states, equilibrium, Poincaré recurrence times, and the like has been achieved through the detailed analysis of systems with small numbers of particles or degrees of freedom. These studies were catalyzed by the now famous 1955, unpublished Los Alamos report of Fermi, Pasta, and Ulam, which is now known as the FPU problem. Excellent reviews can be found in the works of Ford[7] and Weissert[8], and today the FPU problem still represents an active area of research in theoretical/statistical mechanics.

3.6 Reduced Liouville Equation in Generalized Coordinates

Following the same procedures as in Sec. 3.3, we can also write reduced forms of the Liouville equation in generalized coordinates and conjugate momenta. Integrating Eq. (2.2) over (s–u) variables leads to

$$\frac{\partial \rho_u}{\partial t} + \sum_{i=1}^{u} \int\int \left[\frac{\partial H}{\partial p_i}\frac{\partial \rho_s}{\partial q_i} - \frac{\partial H}{\partial q_i}\frac{\partial \rho_s}{\partial p_i}\right] dq^{s-u}\, dp^{s-u}$$

$$+ \sum_{i=u+1}^{s} \int\int \left[\frac{\partial H}{\partial p_i}\frac{\partial \rho_s}{\partial q_i} - \frac{\partial H}{\partial q_i}\frac{\partial \rho_s}{\partial p_i}\right] dq^{s-u}\, dp^{s-u} = 0 \qquad (3.58)$$

†See Ref. 6. Subramanyan Chandrasekhar received the 1983 Nobel Prize in Physics for his theoretical studies on the physical processes important to the structure and evolution of stars.

where

$$\rho_u \equiv \int\int \rho_s \, dq^{s-u} dp^{s-u} \tag{3.59}$$

Because of the possible complicated nature of the hamiltonian and associated variables, no further general simplification of Eq. (3.58) can be performed; every particular system must be analyzed separately. This is illustrated in Example 3.2.

Example 3.2 Reduced Liouville Equation for Interacting Dipole It is desired to write the lowest-order version of the contracted Liouville equation for a system of interacting dipoles. An important physical example of this is a polar gas composed of N indistinguishable molecules, such as those listed in Table 1.2.

From our previous examples, Eqs. (1.41) and (2.63) we can immediately write the hamiltonian for N interacting dipoles as

$$\begin{aligned} H = &\sum_{i=1}^{N} \left[\frac{1}{4m} \left(p_{x_i}^2 + p_{y_i}^2 + p_{z_i}^2 \right) + \frac{1}{mb^2} \left(p_{\theta_i}^2 + \frac{1}{\sin^2 \theta_i} p_{\phi_i}^2 \right) \right] \\ &- \frac{1}{2} \sum_i \sum_{\substack{j \\ j \neq i}} \frac{\mu_i}{r_{ij}^3} [2 \cos\theta_i \cos\theta_j - \sin\theta_i \sin\theta_j \cos(\phi_i - \phi_j)] \end{aligned} \tag{3.60}$$

where we have assumed pairwise additivity of the interaction potential. Now, consider the lowest-order equation and define

$$f_1 = f_1(\mathbf{r}_1, \mathbf{p}_1, \theta_1, \phi_1, p_{\theta_1}, p_\phi, t) \tag{3.61}$$

and

$$f_2 = f_2(\mathbf{r}_1, \mathbf{r}_2, \mathbf{p}_1, \mathbf{p}_2, \theta_1, \theta_2, \phi_1, \phi_2, p_{\theta_1}, p_{\theta_2}, p_{\phi_1}, p_{\phi_2}, t) \tag{3.62}$$

The lowest-order Liouville equation in terms of f becomes

$$\begin{aligned} &\frac{\partial f_1}{\partial t} + \frac{1}{(N-1)!} \int\int \left(\frac{\partial H}{\partial p_1} \frac{\partial f_N}{\partial q_1} - \frac{\partial H}{\partial q_1} \frac{\partial f_N}{\partial p_1} \right) dq^{N-1} dp^{N-1} \\ &+ \frac{1}{(N-1)!} \sum_{i=2}^{N} \int\int \left(\frac{\partial H}{\partial p_i} \frac{\partial f_N}{\partial q_i} - \frac{\partial H}{\partial q_i} \frac{\partial f_N}{\partial p_i} \right) dq^{N-1} dp^{N-1} = 0 \end{aligned} \tag{3.63}$$

where $q_i \in (x_i, y_i, z_i, \theta_i, \phi_i)$ and $p_i \in (p_{x_i}, p_{y_i}, p_{z_i}, p_{\theta_i}, p_{\phi_i})$

Using the hamiltonian, Eq. (3.60), the second term on the left-hand side of Eq. (3.63) can be written as

$$
\begin{aligned}
&\frac{1}{(N-1)!}\int\!\!\int\left(\frac{\partial H}{\partial p_1}\frac{\partial f_N}{\partial q_1}-\frac{\partial H}{\partial q_1}\frac{\partial f_N}{\partial p_1}\right)dq^{N-1}dp^{N-1}\\
&=\frac{\partial H}{\partial p_1}\frac{\partial f_1}{\partial q_1}-\int\!\!\int\frac{\partial u}{\partial q_1}\frac{\partial f_2}{\partial p_1}dq_2dp_2\\
&=\frac{1}{2m}\left(p_{x_1}\frac{\partial f_1}{\partial x_1}+p_{y_1}\frac{\partial f_1}{\partial y_1}+p_{z_1}\frac{\partial f_1}{\partial z_1}\right)+\frac{2}{mb^2}\left(p_{\theta_1}\frac{\partial f_1}{\partial \theta_1}+\frac{p_{\phi_1}}{\sin^2\theta_1}\frac{\partial f_1}{\partial \phi_1}\right)\\
&-\int\!\!\int\left(\frac{\partial u}{\partial x_1}\frac{\partial f_2}{\partial p_{x_1}}+\frac{\partial u}{\partial y_1}\frac{\partial f_2}{\partial p_{y_1}}+\frac{\partial u}{\partial z_1}\frac{\partial f_2}{\partial p_{z_1}}+\frac{\partial u}{\partial \theta_1}\frac{\partial f_2}{\partial p_{\theta_1}}+\frac{\partial u}{\partial \phi_1}\frac{\partial f_2}{\partial p_{\phi_1}}\right)dq_2dp_2
\end{aligned}
\tag{3.64}
$$

where

$$u=\frac{\mu}{r^3}[2\cos\theta_1\cos\theta_2-\sin\theta_1\sin\theta_2\cos(\phi_1-\phi_2)] \tag{3.65}$$

$$r=|\mathbf{r}_1-\mathbf{r}_2|,\quad dq_2=dx_2dy_2dz_2d\theta_2d\phi_2,\quad dp_2=dp_{x_2}dp_{y_2}dp_{z_2}dp_{\theta_2}dp_{\phi_2}$$

Now, the third term on the left-hand side of Eq. (3.44) can be easily shown to be zero by application of Gauss' theorem and the properties

$$f_N\to 0\quad\text{as}\quad x_2,y_2,z_2\to\pm\infty \tag{3.66}$$

$$
\begin{aligned}
\frac{\partial H}{\partial q_2}f_N\to 0\quad\text{as}\quad p_2\to\pm\infty,\ &[q_2\in(x_2,y_2,z_2,\theta_2,\phi_2)]\\
&[p_2\in(p_{x_2},p_{y_2},p_{z_2},p_{\theta_2},p_{\phi_2})]
\end{aligned}
\tag{3.67}
$$

and the symmetry conditions

$$f_N(\theta_2=\pi)=f_N(\theta_2=-\pi) \tag{3.68}$$

$$f_N(\phi_2=0)=f_N(\phi_2=2\pi) \tag{3.69}$$

Thus, we are left with the most general form of the Boltzmann transport equation for a polar gas as

$$
\begin{aligned}
&\frac{\partial f_1}{\partial t}+\frac{1}{2m}\left(p_{x_1}\frac{\partial f_1}{\partial x_1}+p_{y_1}\frac{\partial f_1}{\partial y_1}+p_{z_1}\frac{\partial f_1}{\partial z_1}\right)+\frac{2}{mb^2}\left(p_{\theta_1}\frac{\partial f_1}{\partial \theta_1}+\frac{p_{\phi_1}}{\sin^2\theta_1}\frac{\partial f_1}{\partial \phi_1}\right)\\
&=\int\!\!\int\left(\frac{\partial u}{\partial x_1}\frac{\partial f_2}{\partial p_{x_1}}+\frac{\partial u}{\partial y_1}\frac{\partial f_2}{\partial p_{y_1}}+\frac{\partial u}{\partial z_1}\frac{\partial f_2}{\partial p_{z_1}}+\frac{\partial u}{\partial \theta_1}\frac{\partial f_2}{\partial p_{x_1}}+\frac{\partial u}{\partial \phi_1}\frac{\partial f_2}{\partial p_{\phi_1}}\right)dq_2dp_2
\end{aligned}
\tag{3.70}
$$

Following the previous analysis of the Boltzmann transport equation (Sec. 3.3), the integral term can be expressed in terms of the precollisional and postcollisional velocities. This is left to the reader as an exercise (Prob. 3.9).

3.7 Summary

All of us make decisions about what is important or not important in our daily activities. So too do molecules of substances. They move about under the local influence of surrounding molecules and external field forces. Distant molecules have little influence on their dynamic behavior. The thermodynamic and transport property expressions to be developed in subsequent chapters will be based on molecular probability density functions, involving only a small group of interacting molecules. In this chapter we have laid the foundation for the determination of these reduced probability density functions. The basic equation describing their behavior is the so-called reduced Liouville equation (RLE). We have shown that the lowest nontrivial form of the RLE involves just two interacting molecules. This lowest-ordered RLE leads to the famous Boltzmann transport equation, which has a rich history in the field of gas kinetic molecular theory. It also leads to the law of entropy increase and to the expected molecular equilibrium distribution of gases (Maxwell's distribution). In the case of dense gases, liquids, and solids, the close molecular packing requires a higher-ordered RLE involving the interactions of several or more molecules. Thus, theoretically, the statistical mechanical analysis of dense gases, liquids, and solids is much more mathematically difficult and still remains a fruitful area of research. We have shown the particular forms of the RLE that must be considered for these more complex systems, involving not only more dimensions or molecules, but also different degrees of freedom associated with rotational modes of motion in addition to translational ones. We now wish to turn to the specific determination of probability density functions and the expressions for the equilibrium thermodynamic and nonequilibrium transport properties. We will first look at the equilibrium properties in the next chapter, and then we will turn our attention to the less developed nonequilibrium or transport properties in the final two chapters.

Problems

3.1 Prove Eq. (3.6).

3.2 Prove Eq. (3.12).

3.3 What are the odds of matching three out of six numbers in Example 3.1? (Ans. 1 in 153).

3.4 Show that Eq. (3.24) follows from Eq. (3.23).

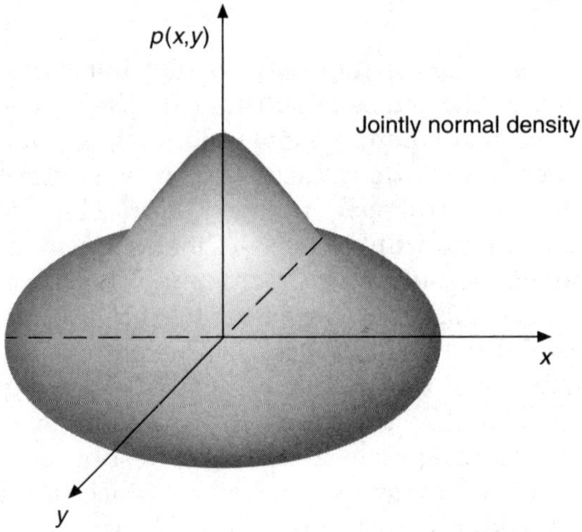

Figure 3.6 Jointly normal random density function.

3.5 Write the reduced Liouville equation for $s = 2$. Truncate the interactions to triplets only, i.e., set $f_4 = 0$. What is the differential equation for f_3?

3.6 Two random variables x and y are said to be *jointly normal* if their probability density function is given by

$$p(x, y) = \frac{1}{2\pi\sigma_1\sigma_2\sqrt{1-\alpha^2}} \times \exp\left\{-\frac{1}{2\sqrt{1-\alpha^2}}\left[\frac{(x-\bar{x})^2}{\sigma_1^2} - \frac{2\alpha(x-\bar{x})(y-\bar{y})}{\sigma_1\sigma_2} + \frac{(y-\bar{y})^2}{\sigma_2^2}\right]\right\}$$

where σ_1, σ_2, $\bar{x}$, $\bar{y}$, and α are constants (see Fig. 3.6). Show that the contracted probability densities $p(x)$ and $p(y)$ are given by the normal density functions

$$p(x) = \frac{1}{\sqrt{2\pi}\sigma_1} \exp\left[-\frac{1}{\sigma_1^2}(x-\bar{x})^2\right]$$

and

$$p(y) = \frac{1}{\sqrt{2\pi}\sigma_2} \exp\left[-\frac{1}{2\sigma_2^2}(y-\bar{y})^2\right]$$

Thus, σ_1 and σ_2 are the standard deviations and $\bar{x}$ and $\bar{y}$ are the mean values of the density functions $p(x)$ and $p(y)$, respectively. The parameter α appearing in the expression $p(x, y)$ is called the correlation coefficient. If $\alpha = 0$, x and y are independent random variables and, thus,

$$p(x, y) = p(x)p(y)$$

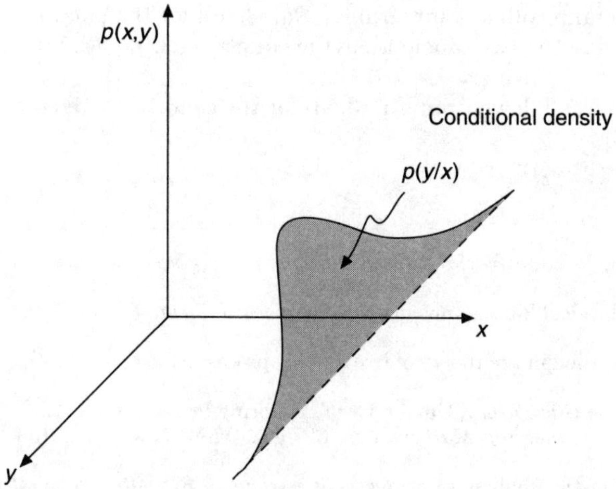

Figure 3.7 Condition density function.

3.7 Conditional probability density functions are often used in statistical mechanics. A conditional density function is defined as

$$p(y|x) = \frac{p(x, y)}{p(x)}$$

and

$$p(x|y) = \frac{p(x, y)}{p(y)}$$

The conditional probability density $p(y|x)$ represents the probability of finding y given a particular x. Show that for the jointly normal density function

$$p(y|x) = \frac{1}{\sigma_2\sqrt{2\pi(1-\alpha^2)}} \exp\left\{-\frac{1}{2\sigma_2^2(1-\alpha^2)}\left[(y-\bar{y}) - \frac{\alpha\sigma_2}{\sigma_1}(x-\bar{x})\right]^2\right\}$$

The conditional density function is shown in Fig. 3.7.

3.8 Show that the characteristic equations of Eq. (3.29) lead to the conservation laws

$$\mathbf{p}_1 + \mathbf{p}_2 = C_1$$

$$\frac{1}{2m_1}(\mathbf{p}_1 \cdot \mathbf{p}_1) + \frac{1}{2m_2}(\mathbf{p}_2 \cdot \mathbf{p}_2) + \phi(|\mathbf{r}_{12}|) = C_3$$

where

$$\phi(|\mathbf{r}_{12}|) \equiv \int \mathbf{F}_{12} \cdot d\mathbf{r}_{21}$$

How could the solution methods for linear PDE's developed in Chap. 2 be used to obtain a more general form of the Boltzmann collision integral?

3.9 Derive the Boltzmann collision integral [cf. Eq. (3.36)] for the polar gas. Begin by simplifying the two-dipole Liouville equation [cf. Eq. (3.29)].

3.10 Prove that Eq. (3.57) follows from Eq. (3.53) for the conditions given by Eqs. (3.54)–(3.56).

References

1. A. Papoulis, *Probability, Random Variables and Stochastic Processes*, McGraw-Hill, New York, 1965.
2. F. C. Andrews, On the classical Boltzmann equation for gases, *J. Chem. Phys.*, **35**, 922, 1962.
3. E. A. Frieman, A new method in the theory of irreversible processes, *J. Math. Phys*, **4**, 410, 1963.
4. L. Boltzmann, *Lectures on Gas Theory*, University of California Press, CA, 1964.
5. E. Kreyszig, *Advanced Engineering Mathematics*, 6th ed., Wiley, New York, 1988, p. 524.
6. S. Chandrasekhar, Stochastic problems in physics and astronomy, *Rev. Mod. Phys.*, **15**, 1, 1943.
7. J. Ford, The Fermi-Pasta-Ulam problem: Paradox turns discovery, *Physics Reports*, **213**, 271, 1992.
8. T. P. Weissert, *The Genesis of Simulation Dynamics. Pursuing the Fermi-Pasta-Ulam Problem*, Springer-Verlag, New York, 1997.

Further Reading

J. O. Hirschfelder, C. F. Curtiss, and R. B. Bird, *The Molecular Theory of Gases and Liquids*, Wiley, New York, 1960. [A classic text on which most of the discussion of reduced density functions given here was based.]

S. Chapman and T. J. Cowling, *The Mathematical Theory of Non-Uniform Gases*, 3rd ed., Cambridge University Press, Cambridge, 1970. [This book extensively covers solutions to the Boltzmann transport equation. Expressions for the viscosity, thermal conductivity, and diffusivity of dilute gases are given for various types of intermolecular potentials.]

Kinetic molecular theory of gases

E. G. D. Cohen, Kinetic theory: Understanding nature through collisions, *Am. J. Physics*, **61**, 524, 1993.

C. Cercignani, *The Boltzmann Equation and Its Applications*, Springer-Verlag, New York, 1988.

J. R. Dorfman and H. van Beijeren, The kinetic theory of gases, in B. J. Berman, (ed.), *Statistical Mechanics*, Part B, Plenum Press, New York, 1977.

S. Harris, *An Introduction to the Theory of the Boltzmann Equation*, Holt, Rinehart & Winston, New York, 1971.

Chapter

4

Equilibrium Solution to the Liouville Equation and the Thermodynamic Properties of Matter

"And I'll prove to you," yelled the South-Going Zax,
That I can stand here in the prairie of Prax
For fifty-nine years! For I live by a rule
That I learned as a boy back in South-Going School.
Never budge! That's my rule. Never budge in the least!
Not an inch to the west! Not an inch to the east!
I'll stay here, not budging! I can and I will
If it makes you and me and the whole world stand still!"

– THEODOR SEUSS GEISEL,
The Zax

4.1 Introduction to the Equilibrium Behavior of Matter

In the introductory chapter we clearly distinguished between so-called global and local equilibrium states. As we know, in global equilibrium states there are no spatial or temporal variations in the system density, temperature, pressure, entropy, and the like. How are such states arrived at when we know that matter is composed of molecules in constant random or thermal motion with continuous intermolecular interactions and dynamics? Our goal in this chapter is to see how such states theoretically come about and to introduce formal, molecular-based definitions of temperature, pressure, entropy, and internal energy. We will see that these familiar macroscopic quantities represent particular averages over the molecular properties of velocity, kinetic energy, and force interactions. When these average quantities remain spatially and temporally invariant throughout the system, a global equilibrium state is achieved.

Following the framework of this text, we present a derivation of the equilibrium probability density functions of matter beginning directly with the Liouville equation. As will be evident in subsequent chapters, this derivation will allow for a harmonious study of both equilibrium and nonequilibrium systems within the framework of the Liouville equation or reduced Liouville equation for classical systems. Fundamental aspects of equilibrium systems, including time invariance, isotropy, and normalization, are used explicitly in the development of the equilibrium solution to the Liouville equation. The solution presented here uses a straightforward separation of variables technique leading to a noneigenvalue problem. The equilibrium solution is used to construct all "basis" thermodynamic functions for gases and liquids. It is shown that all basis thermodynamic functions for "dilute" gases, can be obtained explicitly from knowledge of the two-body intermolecular interaction potential. Integral equations for configurational density functions necessary in the treatment of dense gases and liquids are also shown to follow straightforwardly from the equilibrium solution to the Liouville equation or the reduced Liouville equation. Some specific chemical systems are selected for sample calculations of thermodynamic functions. We note that we will not use the so-called *most probable distribution* methods† to obtain the classical equilibrium behavior of matter. Rather, we will obtain the equilibrium behavior directly from the Liouville equation itself. This will prove to be extremely useful in Chap. 6 when the "closure" problem of the transport (nonequilibrium) equations are treated. We summarize this chapter by including a discussion of questions concerning the uniqueness and existence of equilibrium states, which still remains an area of important theoretical research.

4.2 Solution to the Liouville Equation under Equilibrium Conditions

For a system of molecules at equilibrium, the density function f_N is independent of time and the Liouville equation becomes

$$\sum_{k=1}^{N} \frac{\partial}{\partial \mathbf{r}_k} \cdot \left(\frac{\mathbf{p}_k}{m} f_N \right) + \sum_{k=1}^{N} \frac{\partial}{\partial \mathbf{p}_k} \cdot (\mathbf{F}_k f_N) = 0 \tag{4.1}$$

†See the Further Reading section at the end of this chapter.

Now, we assume a separable solution and that the force is configurational dependent only $[\mathbf{F}_k(\mathbf{r}^N)]$. Thus, we seek a separable, stationary solution to Eq. (4.1) as

$$f_N(\mathbf{r}^N, \mathbf{p}^N) = n_N(\mathbf{r}^N)h_N(\mathbf{p}^N) \tag{4.2}$$

giving

$$\sum_k \left(\frac{\mathbf{p}_k}{m} h_N \cdot \frac{\partial n_N}{\partial \mathbf{r}_k} + \mathbf{F}_k \cdot n_N \frac{\partial h_N}{\partial \mathbf{p}_k} \right) = 0 \tag{4.3}$$

For a system of point particles at equilibrium, the function n_N depends only on the separation distances between particles. Functionally, we must have

$$n_N = n_N(\{r_{ij}\} \qquad i = 1, 3N, j < i) \tag{4.4}$$

where $r_{ij} = |\mathbf{r}_j - \mathbf{r}_i|$ is the separation distance between any two particles (note that $r_{ij} = r_{ji}$, and $\{r_{ij}\}$ denotes the complete set of relative coordinates). Using the chain rule for partial derivatives, it also follows from Eq. (4.4) that

$$\frac{\partial n_N}{\partial \mathbf{r}_k} = \sum_{\substack{i \\ i \neq k}} \frac{\partial n_N}{\partial \mathbf{r}_{ik}} \tag{4.5}$$

In this section, for the sake of simplicity, we will assume that the intermolecular potential function for point particles depends only on the separation distances, i.e.,

$$\Phi(\mathbf{r}^N) = \Phi(\{r_{ij}\} \qquad i = 1, 3N, j < i) \tag{4.6}$$

and, again according to the chain rule

$$\mathbf{F}_k = -\frac{\partial \Phi(\mathbf{r}^N)}{\partial \mathbf{r}_k} = -\sum_{\substack{i \\ i \neq k}} \frac{\partial \Phi}{\partial \mathbf{r}_{ik}} \tag{4.7}$$

Thus, it is seen that both $\partial n_N/\partial \mathbf{r}_k$ and $\partial \Phi/\partial \mathbf{r}_k$ functionally represent the *sum of all* gradients in the functions n_N and Φ, respectively, in terms of the pair-particle separation distances from a given locator molecule at $\mathbf{r}_k$. Because there are no spatial variations in any thermodynamic property for a system at equilibrium, these terms must be independent of the choice of the locator molecule vector $\mathbf{r}_k$ with momentum $\mathbf{p}_k$. Thus, each term of the sum in Eq. (4.3) must be identical, and since the sum of the terms is zero, each term must be identically zero. For any kth molecule, dividing through Eq. (4.3) by $h_N n_N$ gives

$$\frac{\mathbf{p}_k}{m}\frac{1}{n_N}\cdot\frac{\partial n_N}{\partial \mathbf{r}_k}+\mathbf{F}_k\cdot\frac{1}{h_N}\frac{\partial h_N}{\partial \mathbf{p}_k}=0 \tag{4.8}$$

For an isotropic system where there is no preferred direction, the component behavior is given accordingly by

$$\frac{p_{k_l}}{m}\frac{1}{n_N}\frac{\partial n_N}{\partial r_{k_l}}+F_{k_l}\frac{1}{h_N}\frac{\partial h_N}{\partial p_{k_l}}=0 \tag{4.9}$$

Dividing through this equation by $F_{k_l}p_{k_l}/m$ gives,

$$\frac{1}{n_N F_{k_l}(\mathbf{r}^N)}\frac{\partial n_N}{\partial r_{k_l}}=-\frac{m}{h_N p_{k_l}}\frac{\partial h_N}{\partial p_{k_l}}=\lambda \tag{4.10}$$

where λ is a separation constant.

Let's first look at the momentum space part. For the momentum space function h_N we have

$$\frac{\partial h_N}{\partial p_{k_l}}+\frac{\lambda}{m}h_N p_{k_l}=0 \qquad k=1,N \qquad l=1,3 \tag{4.11}$$

Integrating gives,

$$h_N=c_j\exp\left(-\frac{\lambda' p_j^2}{2}\right) \tag{4.12}$$

where $c_j(p^{3N-j})$ is an arbitrary function of all momentum coordinates except p_j. Thus, since Eq. (4.12) holds for any j ($j=1,3N$), the general solution to Eq. (4.11) must be

$$h_N=d\prod_j\exp\left(-\frac{\lambda' p_j^2}{2}\right)=d\exp\left[-\sum_j\left(\frac{\lambda' p_j^2}{2}\right)\right] \tag{4.13}$$

where d is an arbitrary constant. The following conditions exist on f_N:

(i)

$$\int f_N d\mathbf{p}^N=n_N\Rightarrow\int h_N d\mathbf{p}^N=1 \qquad \text{(normalization)} \tag{4.14}$$

(ii)

$$\int \mathbf{p}_k f_N d\mathbf{p}^N=0\Rightarrow\int \mathbf{p}_k h_N d\mathbf{p}^N=0 \qquad \text{(no net momentum)} \tag{4.15}$$

(iii)

$$\frac{1}{2m}\int p_k^2 f_N d\mathbf{p}^N=\frac{3}{2}n_N kT \qquad \text{(kinetic theory temperature)} \tag{4.16}$$

From condition (i), we have

$$d \prod_j \int_{-\infty}^{\infty} \exp\left(-\frac{\lambda' p_j^2}{2}\right) dp_j = 1 \tag{4.17}$$

or

$$d \prod_j \left[\left(\frac{2\pi}{\lambda'}\right)^{1/2}\right] = 1 \Rightarrow d = (\lambda'/2\pi)^{3N/2} \tag{4.18}$$

Note that condition (ii) is satisfied by inspection (antisymmetric integrand).

From condition (iii), which generally defines temperature in the classical sense [cf. Eq. (3.40)], we have

$$\frac{1}{2m}\int p_k^2 h_N d\mathbf{p}^N = \frac{3}{2}kT \tag{4.19}$$

Noting that $p_k^2 = p_{k_x}^2 + p_{k_y}^2 + p_{k_z}^2$ and, thus, for isotropic systems any component of Eq. (4.19) becomes

$$\frac{1}{2m}\int p_j^2 h_N d\mathbf{p}^N = \frac{1}{2}kT \tag{4.20}$$

i.e., each component contributes $\frac{1}{2}kT$ to the average kinetic energy.

Substituting for h_N in Eq. (4.20) gives

$$d \prod_j \int_{-\infty}^{\infty} p_j^2 \exp\left[-\frac{\lambda' p_j^2}{2}\right] dp_j = mkT \tag{4.21}$$

and integrating leads to

$$d \prod_j \left(\frac{\pi^{1/2}2^{1/2}}{\lambda'^{3/2}}\right) = mkT \Rightarrow \lambda' = \frac{1}{mkT} \Rightarrow \lambda = \frac{1}{kT} \tag{4.22}$$

Summarizing our results for the momentum space function, we have

$$h_N = \frac{1}{[2\pi mkT\,]^{3N/2}} \exp\left\{-\sum_{j=1}^{3N} \frac{p_j^2}{2mkT}\right\} \tag{4.23}$$

Now, we turn our attention to the configurational part. The equilibrium N-particle configurational density function is obtained from Eq. (4.10) as

$$kT\frac{\partial \ln[n_N(\mathbf{r}^N)]}{\partial \mathbf{r}_i} = \frac{\partial}{\partial \mathbf{r}_i}\Phi(\mathbf{r}_N) \tag{4.24}$$

Integrating gives

$$n_N(\mathbf{r}^N) = c_1 \exp[-\Phi(\mathbf{r}^N)/kT] \tag{4.25}$$

where c_1 is an integration constant that follows from the normalization condition [Eq. (3.10)]

$$\int n_N(\mathbf{r}^N) d\mathbf{r}^N = N! \tag{4.26}$$

leading to

$$c_1 = \frac{N!}{\int \exp[-\Phi(\mathbf{r}^N)/kT\,] d\mathbf{r}^N} \tag{4.27}$$

Thus, the N-particle configurational distribution function can be expressed as

$$n_N(\mathbf{r}^N) = \frac{W_N(\mathbf{r}^N)}{Q_N} \tag{4.28}$$

where $W_N(\mathbf{r}^N)$ is called the Boltzmann factor

$$W_N(\mathbf{r}^N) \equiv \exp[-\Phi(\mathbf{r}^N)/kT] \tag{4.29}$$

and Q_N is called the configurational integral

$$Q_N \equiv \frac{1}{N!} \int \exp[-\Phi(\mathbf{r}^N)/kT] d\mathbf{r}^N \tag{4.30}$$

For an ideal gas, $\Phi(\mathbf{r}^N) = 0$ and $Q_N = V^N/N!$ Note that the configurational integral is proportional to the so-called partition function introduced in most probable distribution methods,† as shown next.

The partition function is defined by

$$Z_N \equiv [N! h^{3N}]^{-1} \int \int \exp\left\{ -\frac{1}{kT} H(\mathbf{r}^N, \mathbf{p}^N) \right\} d\mathbf{r}^N d\mathbf{p}^N \tag{4.31}$$

where h is Planck's constant and the hamiltonian H is given in cartesian coordinates by

$$H = \sum_{j=1}^{3N} \frac{1}{2} \frac{p_j^2}{m} + \Phi(\mathbf{r}^N) \tag{4.32}$$

†See the Further Reading section at the end of this chapter.

The momentum integrations in Eq. (4.31) can be carried out straightforwardly and, using Eq. (4.30), these lead to the relationship between the partition function and the configurational integral as (Prob. 4.1)

$$Z_N = \frac{1}{h^{3N}} [2\pi mkT]^{3N/2} Q_N \tag{4.33}$$

Summarizing, from Eqs. (4.23) and (4.28) through (4.30), the equilibrium solution to the Liouville equation in terms of the configurational integral is

$$f_N = \frac{1}{[2\pi mkT]^{3N/2}} \exp\left\{ -\sum_{j=1}^{3N} \frac{p_j^2}{2mkT} \right\} \frac{W_N(\mathbf{r}^N)}{Q_N} \tag{4.34}$$

Also, note that this solution corresponds to the so-called *canonical ensemble*, where N, V, and T are held constant over the ensemble [cf. Eq. (4.16)].† The canonical ensemble is illustrated in Fig. 4.1.

Interlude 4.1 Physical Basis of the Canonical and Microcanonical Ensemble We have seen that a canonical ensemble is one in which the number of molecules N volume V and temperature T over the ensemble are all held at some fixed values. The thermodynamic properties are derived as functions of these variables; specifically, $S = S(N, V, T)$, $P = P(N, V, T)$, and $U = U(N, V, T)$.

Physically, how could we obtain such an ensemble? First, consider the laboratory "temperature bath", sketched in Fig. 4.2. Experimentally, the system of interest is placed in the temperature bath where it can exchange energy with the bath. Now, if the bath is infinite in extent (infinite reservoir), the bath temperature remains constant.‡ The so-called *zeroth law of thermodynamics* states that two systems in thermal contact with each other (i.e., energy flows freely between the two systems) will have the same temperature at equilibrium (i.e., at long contact times); thus, $T_b = T_s$ if the contact time is sufficiently large. After a sufficiently long time, when the bath temperature and system temperature have equilibrated, the thermodynamic properties

†The so-called *microcanonical* ensemble is characterized by constants N, V, and E, where E is the total energy or hamiltonian. For the microcanonical ensemble, the equilibrium solution to the Liouville equation is simply f_N = constant (For more details, see the Further Reading section at the end of this chapter). The microcanonical ensemble is illustrated in Fig. 4.3.

‡Practically speaking, since we cannot construct an infinite reservoir, we attempt to maintain a constant temperature in the bath by using a controller that either adds energy or takes energy away using a heating or cooling source, respectively. For our arguments here, we will ignore these additional practical complexities.

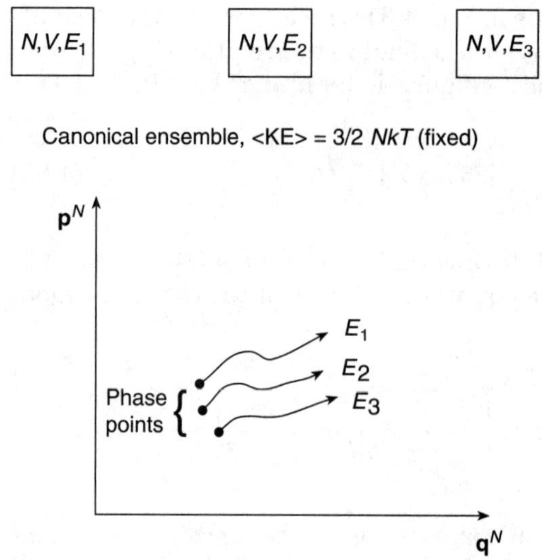

Figure 4.1 Illustration of the canonical ensemble. Each phase point of the ensemble moves along a constant energy (isoenergetic) surface. (Only three phase points are shown.) The average kinetic energy of the molecules over the ensemble is fixed at a prespecified value in the canonical ensemble.

of the system, such as the pressure, are measured. Let's look at the process more carefully from the molecular level.

In the *absence of any external energy transfer*, we should recall that when molecules of the system interact with one another, they do so *without a net*

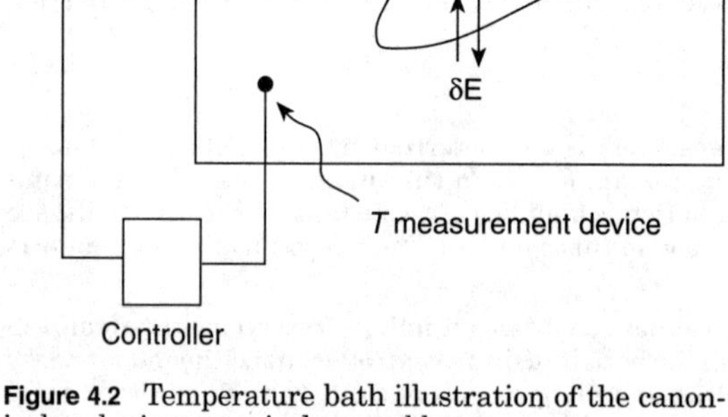

Figure 4.2 Temperature bath illustration of the canonical and microcanonical ensemble.

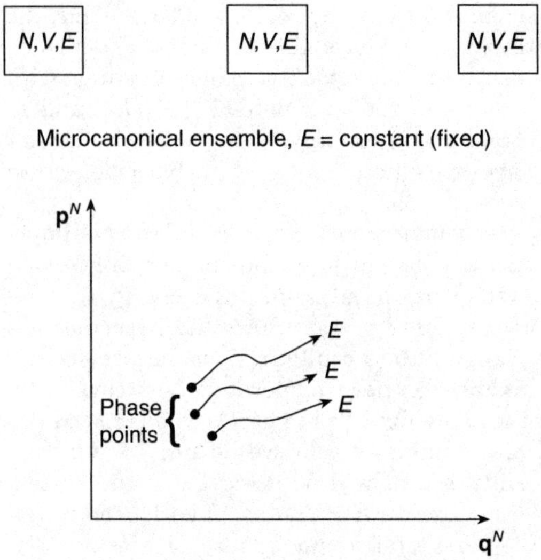

Figure 4.3 Illustration of the microcanonical ensemble. Each phase point of the ensemble moves along the same isoenergetic surface. (Only three phase points are shown.) The total energy is prespecified in the microcanonical ensemble; the temperature or average kinetic energy is not known ahead of time.

gain or loss of total energy. The total energy comprises the kinetic energies of the molecules and the potential energies of intermolecular interactions. The kinetic energy and intermolecular interaction energy of the system continually change as molecules move, interact, and change configurations, but the total energy must remain constant in the absence of any external energy transfer. In our temperature bath system, however, system molecules near the bath-system interface continuously exchange energy with the bath molecules usually via a thermally conducting wall. It is this exchange of energy that maintains $T_b = T_s$.

Now, consider an approximate description where the energy transfer between the bath and the system is done discretely at the end of small time intervals. Over each small, discrete time period between the addition or subtraction of the system energy with the bath, the system molecules move along a constant energy surface. This "isoenergetic" period corresponds to the trajectory of *one* phase point as sketched in Fig. 4.1. Note that the average kinetic energy (and hence, temperature) of the system molecules will change slightly over this period due to changes in molecular configurations. After the addition or withdrawal of the system energy from interaction with the bath, the system molecules move along a new isoenergetic surface giving a new phase point trajectory (Fig. 4.1). In this situation then, both the total energy of the bath and the total energy of the system fluctuate with time. The trajectory time average, as the system moves along various isoenergetic paths, is taken to be equivalent to the ensemble average, and this equivalence is known as the *ergodic hypothesis*. Thus, the total energy of the system in the canonical ensemble fluctuates about some average, equilibrium value—the internal energy U.† Because of contact with the bath, the different energy levels in the ensemble (E_1, E_2, E_3, and so on in Fig. 4.2) result in the average kinetic energy of the system molecules over the ensemble equal to the average kinetic energy of the bath molecules: $T_s = T_b$.

Physically, the microcanonical ensemble (constants E, N, V) can be simply achieved by thermally insulating the system from the surroundings using an adiabatic enclosure. The system begins at some total energy and remains at that total energy for all time (Fig. 4.2). Since the molecules are moving about and constantly changing their configurations, intermolecular interaction energies and the kinetic energies (temperature) are constantly changing as well to maintain the same fixed total energy. The physical difficulty with the microcanonical ensemble is that we do not know ahead of time the equilibrium temperature that will be achieved as the system evolves over time to some average temperature determined by the initial total energy, number of molecules, and total volume. Thus, some sort of trial and error procedure is required to achieve some specified average system temperature.‡,[1]

4.3 The Thermodynamic Functions $\hat{U}$, P, and $\hat{S}$

The "basis" thermodynamic functions for pure (single-component) systems are the specific volume ($\hat{V} \equiv V\tilde{N}/N$), temperature ($T$), specific internal energy ($\hat{U} \equiv U\tilde{N}/N$), pressure ($P$), and specific entropy ($\hat{S} \equiv S\tilde{N}/N$). Note that $\tilde{N}$ is Avogadro's number and N is the total number of molecules. All other thermodynamic functions can be expressed in terms of these basis functions; for example, $\hat{H}$ (enthalpy) $\equiv \hat{U} + P\hat{V}$, $\hat{G}$ (Gibbs free energy) $\equiv \hat{H} - T\hat{S}$, $\hat{A}$ (Helmholtz free energy) $\equiv \hat{U} - T\hat{S}$. Also, the specific heat capacities are defined by $\hat{C}_V \equiv (\partial\hat{U}/\partial T)_{\hat{V}}$ and $\hat{C}_p \equiv (\partial\hat{U}/\partial T)_p$.

†It may seem shocking, but these property fluctuations of equilibrium can tell us something about the nonequilibrium properties of the system. This advanced topic of statistical mechanics is called *time correlation formalism*. (see D. A. McQuarrie in the Further Reading section at end of this chapter.)

‡A derivation of the Sackur-Tetrode formula for the entropy of an ideal gas using the microcanonical ensemble is given by C. Kittel in the Further Reading section at the end of this chapter. Recall that thermodynamic properties are *state variables*, i.e., they only depend on the particular (equilibrium) state (N, V, and T) of the system and not on the path used to obtain that particular state. This means that whatever ensemble is chosen, the resulting functions $S(N, V, T)$ and $U(N, V, T)$ must be the same, i.e., independent of the ensemble type.

By definition, the internal energy for a system at equilibrium is the average total energy in the system, i.e.,

$$U \equiv \frac{\int\int f_N H(\mathbf{r}^N, \mathbf{p}^N)\, d\mathbf{r}^N d\mathbf{p}^N}{\int\int f_N\, d\mathbf{r}^N d\mathbf{p}^N} = \frac{1}{N!}\int\int f_N H(\mathbf{r}^N, \mathbf{p}^N)\, d\mathbf{r}^N d\mathbf{p}^N \tag{4.35}$$

Substituting Eq. (4.30) into Eq. (4.35) and carrying out the momentum integrations (see Prob. 4.1) gives

$$U = \frac{3}{2}NkT + \frac{1}{N!}\int n_N(\mathbf{r}^N)\Phi(\mathbf{r}^N)\, d\mathbf{r}^N \tag{4.36}$$

Alternatively, using Eqs. (4.30) and (4.36) can be reexpressed in terms of the configurational integral as

$$U = \frac{3}{2}NkT + kT^2 \frac{\partial \ln Q_N}{\partial T} \tag{4.37}$$

or, using Eq. (4.33), in terms of the partition function as

$$U = kT^2 \frac{\partial \ln Z_N}{\partial T} \tag{4.38}$$

The (Gibbs) entropy S_G for a classical system is defined in the canonical ensemble as[†,2]

$$\frac{S_G}{k} \equiv -\frac{\int\int f'_N \ln f'_N\, d\mathbf{r}^N d\mathbf{p}^N}{\int\int f'_N\, d\mathbf{r}^N d\mathbf{p}^N} = -\frac{1}{N!h^{3N}}\int\int f'_N \ln f'_N\, d\mathbf{r}^N d\mathbf{p}^N \tag{4.39}$$

where $f'_N \equiv h^{3N} f_N$ is a dimensionless distribution function necessary for the use of the logarithmic function. Note that Planck's constant h has units of momentum × length (called "action") and f_N has units of (length × momentum)$^{-3N}$. The use of Planck's constant for scaling the distribution function gives the correct classical limit of quantum mechanical definitions of entropy.[2]

It is to be noted that the above definition of entropy is identical to statistical measures of information, and more discussion on the definition of entropy will be given in Chap. 5. Substituting the equilibrium solution, Eq. (4.34), into Eq. (4.39) and carrying out the momentum integrations leads to (Prob. 4.10)

$$\frac{S_G}{k} = N \ln\left[\frac{e^{3/2}(2\pi mkT)^{3/2}}{h^3}\right] - \frac{1}{N!}\int n_N \ln n_N\, d\mathbf{r}^N \tag{4.40}$$

[†]In the microcanonical ensemble, entropy is a measure of the total available phase. (See the Further Reading section at the end of this chapter.)

For an ideal gas, $\Phi(\mathbf{r}^N) = 0$ and $n_N = N!/V^N$, the expression reduces to

$$S_{\text{ideal}} = kN \ln \left[\frac{e^{5/2}(2\pi mkT)^{3/2}V}{Nh^3} \right] \tag{4.41}$$

which is known as the Sackur-Tetrode formula.†

Alternatively, using Eqs. (4.27) and (4.40) can be reexpressed in terms of the configurational integral as

$$S_G = Nk \ln \left[\frac{(2\pi mkT)^{3/2}}{h^3} \right] + \frac{U}{T} + k \ln Q_N \tag{4.42}$$

or, using Eq. (4.30), in terms of the partition function as

$$S_G = \frac{U}{T} + k \ln Z_N \tag{4.43}$$

In general, pressure is defined as the net momentum flux through an imaginary plane that moves with the local average velocity in the system. Since momentum is a vector quantity and the direction of transfer is described by a vector, pressure is, in general, a tensor quantity. In the next chapter, we will show that the pressure tensor arises naturally in the derivation of the macroscopic momentum conservation equation from the Liouville equation. It is written as

$$\mathbf{P} = \mathbf{P}_k + \mathbf{P}_\phi \tag{4.44}$$

where the kinetic part of the pressure tensor $\mathbf{P}_k$ is given by

$$\mathbf{P}_k = m \int \left(\frac{\mathbf{p}_1}{m} - \mathbf{v}_0 \right)^2 f_1(\mathbf{r}_1, \mathbf{p}_1, t)\, d\mathbf{p}_1 \tag{4.45}$$

where $\mathbf{v}_0$ is the local average velocity. The potential part $\mathbf{P}_\phi$ is given by

$$\mathbf{P}_\phi = -\frac{1}{2} \int n_2(\mathbf{r}_1, \mathbf{r}_2, t) \frac{\mathbf{r}_{12}^2}{r_{12}} \frac{d\phi(r_{12})}{dr_{12}} d\mathbf{r}_{12} \tag{4.46}$$

where $\mathbf{r}_{12} = \mathbf{r}_2 - \mathbf{r}_1$ and $r_{12} = |\mathbf{r}_{12}|$. In writing Eq. (4.46), pairwise additivity of the interaction forces has been assumed. Also, $n_2(\mathbf{r}_1, \mathbf{r}_2, t)$

†Note that the classical expression for entropy does not recover the third law of thermodynamics, i.e., $\lim_{T\to 0} S = 0$. However, the corresponding quantum mechanical formula for entropy, which reduces to the classical formula in the "correspondence limit," does indeed lead to the third law of thermodynamics [see pp. 109 and 110, Eq. (2.4–20) of Ref. 2]. From a practical point of view, quantum corrections are usually not important for temperatures away from absolute zero. The exceptions to this are the "quantum" gases He and H_2 (Ref. 2, p. 164, Fig. 3.6–1).

is a two-particle configurational distribution function defined generally by [Eq. (3.11)]

$$n_2(\mathbf{r}_1, \mathbf{r}_2, t) = \frac{1}{(N-2)!} \int n_N(\mathbf{r}^N, t)\, d\mathbf{r}^{N-2} \tag{4.47}$$

We next examine specific forms of the thermodynamic basis functions $(U, S_G, \mathbf{P})$ under equilibrium conditions.†

4.4 Determination of the Configurational Distribution Functions

In order to compute the thermodynamic properties, the many-bodied configurational distribution $n_N(\mathbf{r}^N)$ is needed. An approximate method of treating many-bodied interactions is the assumption of *pairwise additivity*, i.e., the total potential is written as

$$\Phi(\mathbf{r}^N) = \frac{1}{2} \sum_i \sum_{\substack{j \\ j \neq i}} \phi(r_{ij}) = \sum_i \sum_{j<i} \phi(r_{ij}) \tag{4.48}$$

where $\phi(r_{ij})$ is the pair interaction potential between molecule (i) and molecule (j) in the absence of all other molecules. Although the above expression is exact for some interaction types, such as coulombic electrostatic interactions, it is only approximate for other common interactions such as van der Waals interactions.[3]

Under the pairwise additivity assumption, the configurational part of the equilibrium solution to the Liouville equation, Eq. (4.24), can be easily reduced to the following integral equation (Prob. 4.12)

$$kT \frac{\partial \ln n_h}{\partial \mathbf{r}_k} + \frac{\partial \Phi(\mathbf{r}^h)}{\partial \mathbf{r}_k} = -\int \frac{n_{h+1}}{n_h} \frac{\partial \Phi(\mathbf{r}_k, \mathbf{r}_{h+1})}{\partial \mathbf{r}_k}\, d\mathbf{r}_{h+1} \tag{4.49}$$

where

$$\Phi(\mathbf{r}^h) = \frac{1}{2} \sum_{i=1}^{h} \sum_{\substack{j=1 \\ j \neq i}}^{h} \Phi(\mathbf{r}_{ij}) = \sum_{i=1}^{h} \sum_{\substack{j=1 \\ j<i}}^{h} \Phi(\mathbf{r}_{ij}) \tag{4.50}$$

---

†A "back-door" method of obtaining the equilibrium scalar pressure p ($\mathbf{P} = p\mathbf{I}$) is to consider the closed system energy balance for reversible processes with volume as the only mechanical parameter, i.e., $dU = T\,dS - p\,dV$. Thus, $p = -(\partial U/\partial V)_T + T(\partial S/\partial V)_T$; or using Eq. (4.43), in terms of the partition function, $p = kT(\partial \ln Z_N/\partial V)_T$. The energy and entropy balance equations will be obtained in Chaps. 5 and 6.

and

$$n_h = \frac{1}{(N-h)!} \int n_N \, d\mathbf{r}^{N-h} \tag{4.51}$$

is a reduced distribution function for a set of h molecules ($\{h\}$) in the total N molecules; also, k is a member of the set ($\{h\}$). For example, let $h = 2$; then,

$$kT \frac{\partial \ln n_2}{\partial \mathbf{r}_1} + \frac{\partial \Phi(\mathbf{r}_{12})}{\partial \mathbf{r}_1} = -\int \frac{n_3(\mathbf{r}_1, \mathbf{r}_2, \mathbf{r}_3)}{n_2(\mathbf{r}_1, \mathbf{r}_2)} \frac{\partial \Phi(\mathbf{r}_{13})}{\partial \mathbf{r}_1} d\,\mathbf{r}_3 \tag{4.52}$$

where $\mathbf{r}_{12} = \mathbf{r}_2 - \mathbf{r}_1$ and $\mathbf{r}_{13} = \mathbf{r}_3 - \mathbf{r}_1$. Similarly, for $h = 3$, we have

$$kT \frac{\partial \ln n_3}{\partial \mathbf{r}_1} + \frac{\partial}{\partial \mathbf{r}_1}[\Phi(\mathbf{r}_{12}) + \Phi(\mathbf{r}_{23}) + \Phi(\mathbf{r}_{13})] = -\int \frac{n_4}{n_3} \frac{\partial \Phi(\mathbf{r}_{14})}{\partial \mathbf{r}_1} d\,\mathbf{r}_4 \tag{4.53}$$

Equations (4.52) and (4.53) illustrate the BBGKY hierarchy for the equilibrium configurational distribution functions; some truncation (closure) is necessary to obtain solutions. For example, neglecting three-body effects by setting $\Phi(\mathbf{r}_1, \mathbf{r}_3) = 0$, gives from Eq. (4.52) the dilute gas result as

$$n_2^{\mathrm{DG}}(\mathbf{r}_{12}) = n^2 \exp[-\Phi(\mathbf{r}_{12})/kT\,] \tag{4.54}$$

Neglecting four-body effects by setting $\Phi(\mathbf{r}_1, \mathbf{r}_4) = 0$, gives from Eq. (4.53)

$$n_3 = n^3 \exp[-\Phi(\mathbf{r}_{12})/kT\,] \exp[-\Phi(\mathbf{r}_{23})/kT\,] \exp[-\Phi(\mathbf{r}_{13})/kT\,] \tag{4.55}$$

or

$$n_3 = \frac{1}{n^3} n_2^{\mathrm{DG}}(\mathbf{r}_{12}) n_2^{\mathrm{DG}}(\mathbf{r}_{23}) n_2^{\mathrm{DG}}(\mathbf{r}_{13}) \tag{4.56}$$

which is a form of the so-called superposition approximation.[1] A more refined form of the superposition approximation suggested by Eq. (4.56) is[4]

$$n_3 = \frac{1}{n^3} n_2(\mathbf{r}_{12}) n_2(\mathbf{r}_{23}) n_2(\mathbf{r}_{13}) \tag{4.57}$$

where now all pair density functions include three-body effects. Substituting Eq. (4.57) into Eq. (4.52) leads to an integral equation for n_2, originally due to Yvon, Born, and Green.[4,5] Numerical solutions to the integral equation were first obtained by Kirkwood et al.[6] and are considered in more detail in Sec. 4.5. A summary of other more refined closure methods can be found, for example, in the books by McQuarrie and Ziman given in the Further Reading section at the end of this chapter.

In terms of the pairwise additivity assumption, the thermodynamic function U can be obtained by substituting Eq. (4.48) into Eq. (4.36), leading to

$$U = \frac{3}{2}NkT + \frac{1}{2}V \int \Phi(\mathbf{r}_{12})n_2(\mathbf{r}_{12})\,d\mathbf{r}_{12} \tag{4.58}$$

Note that $n_2(\mathbf{r}_{12})$ contains three-body effects via the integral equation for $n_2(\mathbf{r}_{12})$.

Unfortunately, even for pairwise additive systems, it is not possible to directly contract the entropy, Eq. (4.40), solely in terms of the two-particle distribution function, as in the case of the internal energy.[4,7,8] Thus, some sort of expansion process is needed for evaluating the Gibbs entropy as given next.

Let's consider a simple expression for n_N following the development of Eq. (4.55). Extending Eq. (4.55) to the entire N-molecule set, we have

$$n_N \cong \frac{N!}{V^N}\left[\frac{n_2^{\mathrm{DG}}(\mathbf{r}_{12})}{n^2}\cdot\frac{n_2^{\mathrm{DG}}(\mathbf{r}_{13})}{n^2}\cdots\frac{(N)(N-1)}{2}\text{terms}\right] \tag{4.59}$$

where the factor $N!/V^N$ is necessary to satisfy the ideal gas behavior when $\phi(\mathbf{r}_{ij}) = 0$, i.e.,

$$n_N^{\mathrm{ideal}} = N!/V^N \tag{4.60}$$

Note that Eq. (4.60) satisfies the normalization condition: $\int n_N\,d\mathbf{r}^N = N!$. For more refinement, we will drop the superscript DG in Eq. (4.59) as done in Eq. (4.57).

Now, look at the configurational integral in the entropy equation

$$\begin{aligned}&\frac{1}{N!}\int n_N \ln n_N\,d\mathbf{r}^N\\&\cong \frac{1}{N!}\int n_N\{\ln N! - N\ln V - \ln n_2(\mathbf{r}_{12}) - 2\ln n\cdots\}\,d\mathbf{r}^N\end{aligned} \tag{4.61}$$

where we have used Eq. (4.46) to express $\ln n_N$. Carrying out the integrations in Eq. (4.61) including all $(N)(N-1)/2$ terms from Eq. (4.59) gives

$$\begin{aligned}\frac{1}{N!}\int n_N \ln n_N\,d\mathbf{r}^N &= \ln N! - N\ln V - (N)(N-1)\ln n\\&\quad + \frac{(N)(N-1)}{2}\frac{1}{N!}\int n_N \ln n_2(\mathbf{r}_{ij})\,d\mathbf{r}^N\\&= N\ln N - N - N\ln V - (N)(N-1)\ln n\\&\quad + \frac{1}{2}\int\int n_2(\mathbf{r}_{ij})\ln n_2(\mathbf{r}_{ij})\,d\mathbf{r}_i d\mathbf{r}_j\end{aligned} \tag{4.62}$$

where we have used Stirling's approximation: $\ln N! \cong N \ln N - N$, for large N. Note that for an *ideal gas*, $\ln n_2 = 2 \ln n$ and

$$\frac{1}{2}\int\int n_2 \ln n_2 d\mathbf{r}_i d\mathbf{r}_j$$
$$= \ln n \int\int n_2 \, d\mathbf{r}_i \mathbf{r}_j = (N)(N-1)\ln n \quad \text{(ideal gas only)} \tag{4.63}$$

Thus, from Eq. (4.62)

$$\frac{1}{N!}\int n_N \ln n_N \, d\mathbf{r}^N = N \ln n - N \qquad \text{(ideal gas only)} \tag{4.64}$$

Returning to the entropy equation, Eq. (4.40), and using Eq. (4.62) gives

$$\frac{S_G}{k} = N \ln\left[\frac{e^{3/2}(2\pi mkT)^{3/2}}{h^3}\right] + N - N \ln n + (N)(N-1)\ln n$$
$$- \frac{1}{2}\int\int n_2(\mathbf{r}_{ij}) \ln n_2(\mathbf{r}_{ij}) \, d\mathbf{r}_i \, d\mathbf{r}_j$$
$$= S_{\text{ideal}}/k + (N)(N-1)\ln n - \frac{1}{2}\int\int n_2(\mathbf{r}_{ij}) \ln n_2(\mathbf{r}_{ij}) \, d\mathbf{r}_i \, d\mathbf{r}_j \tag{4.65}$$

Alternatively, in terms of the radial distribution function $g_2(\mathbf{r}_{ij})$ defined by

$$g_2(\mathbf{r}_{ij}) \equiv \frac{n_2(\mathbf{r}_{ij})}{n^2} \tag{4.66}$$

and, thus,

$$\ln n_2(\mathbf{r}_{ij}) = 2 \ln n + \ln g_2(\mathbf{r}_{ij}) \tag{4.67}$$

we obtain

$$\frac{S_G}{k} = \frac{S_{\text{ideal}}}{k} - \frac{1}{2}n^2 \int\int g_2(\mathbf{r}_{ij}) \ln g_2(\mathbf{r}_{ij}) \, d\mathbf{r}_i \, d\mathbf{r}_j$$
$$= \frac{S_{\text{ideal}}}{k} - \frac{1}{2}nN \int g_2(\mathbf{r}_{ij}) \ln g_2(\mathbf{r}_{ij}) \, d\mathbf{r}_{ij} \tag{4.68}$$

Equation (4.65) or (4.68), originally due to Green (see Refs. 9 and 10), is the desired result for the entropy in pairwise additive systems. The essential problem then is to determine the two-particle configurational density function from the solution to Eq. (4.52). However, to simply illustrate numerical calculations, we first consider the dilute gas system.

4.5 Thermodynamic Functions for a Dilute Gas

We have shown that for a dilute gas, the equilibrium two-particle configurational density function is given by

$$n_2^{\mathrm{DG}} = \frac{1}{(N-2)!}\int n^N \, d\mathbf{r}^{N-2} = n^2 \exp[-\phi(r_{12})/kT\,] \tag{4.69}$$

Including the momentum space contribution, from Eq. (4.23), the equilibrium solution for the binary (dilute) gas is

$$f_2 = \frac{n^2}{[2\pi mkT\,]^3}\exp\left\{-\sum_{j=1}^{6}\frac{p_j^2}{2mkT}\right\}\exp[-\phi(r_{12})/kT\,] \tag{4.70}$$

Also note that, in general [Eq. (3.7)]

$$f_1(\mathbf{r}_1, \mathbf{p}_1, t) = \frac{1}{(N-1)}\int f_2(\mathbf{r}_1, \mathbf{r}_2, \mathbf{p}_1, \mathbf{p}_2, t)\, d\mathbf{r}_2\, d\mathbf{p}_2 \tag{4.71}$$

Substituting Eq. (4.70) into Eq. (4.71), changing variables from $(\mathbf{r}_1, \mathbf{r}_2)$ to $(\mathbf{r}_1, \mathbf{r}_{12})$, and integrating gives

$$f_1 = \frac{n}{[2\pi mkT\,]^{3/2}}\exp\left\{-\sum_{j=1}^{3}\frac{p_j^2}{2mkT}\right\} \tag{4.72}$$

which is the famous Maxwell-Boltzmann distribution for molecular velocities at equilibrium.

Now let's consider the thermodynamic functions p, U, and S for the dilute gas. Substituting the equilibrium solution, Eq. (4.70), into the pressure equations, Eqs. (4.45) and (4.46), and carrying out the integrations gives[†]

$$\mathbf{P} = p\mathbf{I} \tag{4.73}$$

where

$$p = nkT - \frac{1}{6}n^2\int_0^\infty \exp[-\phi(r_{12})/kT\,]4\pi r_{12}^3\frac{d\phi(r_{12})}{dr_{12}}dr_{12} \tag{4.74}$$

[†]Consider a spherical coordinate system centered at molecule (1). We have that $d\mathbf{r}_{12} = r_{12}^2\sin^2\theta\, d\theta\, d\phi\, dr_{12}$. Also, $(\mathbf{r}_{12}/r_{12}) = \sin\theta\cos\phi\mathbf{e}_x + \sin\theta\sin\phi\mathbf{e}_y + \cos\theta\mathbf{e}_z$, where $\mathbf{e}_x$, $\mathbf{e}_y$, and $\mathbf{e}_z$ are unit vectors in a cartesian coordinate system centered at molecule (1).

The virial equation of state is written as

$$\frac{p\hat{V}}{RT} = 1 + \frac{B(T)}{\hat{V}} + \frac{C(T)}{\hat{V}^2} + \cdots \tag{4.75}$$

where $\hat{V}$ is the molar volume (volume/mole) and $B(T), C(T), \ldots$ are the second, third, ... virial coefficients, respectively, with the temperature dependence shown explicitly.

Now, since

$$n \equiv \tilde{N}/\hat{V} \tag{4.76}$$

and the gas constant R is defined by

$$R \equiv \tilde{N}k \tag{4.77}$$

where $\tilde{N}$ is Avogadro's number, a comparison of Eqs. (4.74) and (4.75) gives an expression for the second virial coefficient as

$$\begin{aligned} B(T) &= -\frac{2\pi\tilde{N}}{3kT}\int_0^\infty \exp[-\phi(r_{12})/kT]\frac{d\phi(r_{12})}{dr_{12}}r_{12}^3\,dr_{12} \\ &= \frac{2\pi\tilde{N}}{3}\int_0^\infty \frac{d}{dr_{12}}\{\exp[-\phi(r_{12})/kT]\}r_{12}^3\,dr_{12} \end{aligned} \tag{4.78}$$

Integrating by parts, this last expression can also be written as

$$B(T) = -2\pi\tilde{N}\int_0^\infty \{\exp[-\phi(r_{12})/kT] - 1\}r_{12}^2\,dr_{12} \tag{4.79}$$

Similarly, the internal energy for the dilute gas can be obtained straightforwardly by substituting Eq. (4.69) into Eq. (4.58) to give

$$\begin{aligned} U = \frac{3}{2}NkT + 2\pi nN\int \phi(r_{12}) \\ \times \exp[-\phi(r_{12})/kT]r_{12}^2\,dr_{12} \qquad \text{(dilute gas only)} \end{aligned} \tag{4.80}$$

It is useful to express this last equation in terms of the temperature derivative of the second virial coefficient as

$$U = \frac{3}{2}NkT - \frac{kT^2N}{\tilde{N}}\left(\frac{\tilde{N}}{\hat{U}}\frac{dB}{dT}\right) \tag{4.81}$$

or, in terms of the specific internal energy ($\hat{U} = U\tilde{N}/N$; energy/mole)

$$\hat{U} = \frac{3}{2}RT - \frac{RT^2}{\tilde{U}}\frac{dB}{dT} \tag{4.82}$$

For the Gibbs entropy, we obtain from Eq. (4.68)

$$S_G = Nk \ln\left[\frac{e^{5/2}(2\pi mkT)^{3/2}}{h^3}\right] - \frac{2\pi kN}{n} \times \int n_2^{\mathrm{DG}}(r_{12}) \ln n_2^{\mathrm{DG}}(r_{12}) r_{12}^2\, dr_{12} \qquad \text{(dilute gas only)} \tag{4.83}$$

The above equation can also be easily rewritten in terms of the so-called radial distribution function $g^{\mathrm{DG}}(r_{12})$ for a dilute gas, i.e.,

$$n_2^{\mathrm{DG}}(\mathbf{r}_{12}) \equiv n^2 g^{\mathrm{DG}}(r_{12}) = n^2 \exp[-\phi(r_{12})/kT] \tag{4.84}$$

giving

$$S_G = S_{\text{ideal}} - 2\pi nkN \int g^{\mathrm{DG}}(r_{12}) \ln g^{\mathrm{DG}}(r_{12}) r_{12}^2\, dr_{12} \tag{4.85}$$

which is in agreement with that given by Green.[9,10] In terms of the temperature derivative of the second virial coefficient, Eq. (4.85) becomes

$$\hat{S}_G = \hat{S}_{\text{ideal}} - \frac{RT}{\hat{V}} \frac{dB}{dT} \tag{4.86}$$

where $\hat{S} = S\tilde{N}/N$.

Thus, the thermodynamic basis functions (P, $\hat{U}$, and $\hat{S}^G$) can be readily determined for dilute gases from Eqs. (4.74), (4.80), and (4.85), respectively, for various interaction potentials such as hard-sphere, point centers of attraction or repulsion, and Lennard-Jones (L-J) interaction potentials. We now determine second virial coefficients for simple intermolecular interactions that admit analytical results.

4.5.1 Hard-sphere interactions

The hard-sphere interaction potential is defined as

$$\phi^{\mathrm{HS}}(r_{12}) = \begin{cases} \infty & r_{12} \le \sigma \\ 0 & r_{12} > \sigma \end{cases} \tag{4.87}$$

where σ is the molecular diameter and thus

$$\exp[-\phi^{\mathrm{HS}}(r_{12})/kT] = U(r_{12} - \sigma) \tag{4.88}$$

where $U(r_{12} - \sigma)$ is the unit step function defined as

$$U(r_{12} - \sigma) = \begin{cases} 0 & r_{12} \le \sigma \\ 1 & r_{12} > \sigma \end{cases} \tag{4.89}$$

The unit step function also has the property

$$\frac{d}{dr_{12}}U(r_{12}-\sigma)=\delta(r_{12}-\sigma) \tag{4.90}$$

where $\delta(r_{12}-\sigma)$ is the so-called Dirac delta function defined as

$$\delta(r_{12}-\sigma)=\begin{cases}0 & r_{12}\neq\sigma\\ \infty & r_{12}=\sigma\end{cases} \tag{4.91}$$

Using the above relationships we readily obtain the hard-sphere second virial coefficient

$$B^{\text{HS, DG}}=\frac{2}{3}\pi\tilde{N}\sigma^3 \qquad \text{(hard sphere, dilute gas)} \tag{4.92}$$

Note that because of the lack of temperature dependence of $B^{\text{HS, DG}}$, we have the intriguing results

$$S^{\text{HS, DG}}=S^{\text{ ideal}} \qquad U^{\text{HS, DG}}=U^{\text{ ideal}} \tag{4.93}$$

4.5.2 Point centers of attraction or repulsion

In this case, the force acting between two particles is given by

$$\mathbf{F}_1=\frac{a}{r_{12}^{e}}\left(\frac{\mathbf{r}_{12}}{r_{12}}\right) \qquad e>1 \tag{4.94}$$

which is the force acting on particle (1) due to a second particle at a distance r_{12} where a and e are constants. The potential associated with this force is determined from

$$-\frac{\partial\phi}{\partial\mathbf{r}_1}=\mathbf{F}_1 \tag{4.95}$$

and thus

$$\phi^{pc}(r_{12})=+\frac{ar_{12}^{-(e-1)}}{(e-1)} \qquad e>1 \tag{4.96}$$

if the constant a is positive, the force is repulsive, and vice versa. Substituting into Eq. (4.79) and integrating gives

$$B(T)=\frac{2\pi\tilde{N}}{3}\left[\frac{kT\,(e-1)}{a}\right]^{3/(e-1)}\Gamma\left(\frac{e-4}{e-1}\right) \qquad e>1 \tag{4.97}$$

Note that the internal energy and entropy for this case can be obtained from Eqs. (4.82) and (4.86) using Eq. (4.97).

Example 4.1 Thermodynamic Properties of Lennard-Jones (6–12) Gases Consider the expression for the second virial coefficient $B(T)$ for the 6–12, Lennard-Jones potential

$$\phi(r_{12}) = 4\epsilon\left[\left(\frac{\sigma}{r_{12}}\right)^{12} - \left(\frac{\sigma}{r_{12}}\right)^{6}\right] \tag{4.98}$$

Substituting Eq. (4.98) into Eq. (4.78) gives

$$B(T) = -\frac{8\pi\tilde{N}\epsilon}{3kT}\int_0^{\infty}\exp\left\{-\frac{4\epsilon}{kT}\left[\left(\frac{\sigma}{r_{12}}\right)^{12} - \left(\frac{\sigma}{r_{12}}\right)^{6}\right]\right\} \times\left[-12\left(\frac{\sigma}{r_{12}}\right)^{12} + 6\left(\frac{\sigma}{r_{12}}\right)^{6}\right] r_{12}^2\,dr_{12} \tag{4.99}$$

Expanding the exponential in a series, the integration in Eq. (4.99) can be carried out analytically leading to (Prob. 4.11)

$$B(T) = \frac{2}{3}\pi\tilde{N}\sigma^3\sum_{j=0}^{\infty}\left(\frac{kT}{\epsilon}\right)^{-(2j+1)/4}\left[-\frac{2^{(j+\frac{1}{2})}}{4j!}\Gamma\left(\frac{2j-1}{4}\right)\right] \tag{4.100}$$

The specific (per mole) thermodynamic functions $\hat{U}$ and $\hat{S}$ require the temperature derivative of $B(T)$, which follows from Eq. (4.100) as

$$\frac{dB}{dT} = \frac{2}{3}\frac{\pi\tilde{N}\sigma^3}{T}\sum_{j=0}^{\infty}\left[-\frac{(2j+1)}{4}\right]\left(\frac{kT}{\epsilon}\right)^{-(2j+1)/4}\left[-\frac{2^{(j+\frac{1}{2})}}{4j!}\Gamma\left(\frac{2j-1}{4}\right)\right] \tag{4.101}$$

Now, let's compare the experimental values of the entropy of methane(g) from Table 4.1 with the theoretical values using the Lennard-Jones potential constants from Table 1.1, i.e., $\epsilon/k = 148.2$ K and $\sigma = 3.817$ Å. We have

$$\hat{S} = \hat{S}_{\text{ideal}} - \frac{RT}{\hat{V}}\frac{dB}{dT} \tag{4.102}$$

TABLE 4.1 Thermodynamic Properties of Methane [p = 0.101325 MPa (1 atm)]

T(K)	$\hat{V}$ (m^3/mol)	$\hat{U}$(J/mol)	$\hat{S}$(J/mol K)	$\hat{C}_V$ (J/mol K)	$\hat{S}_{\text{ideal}} + \hat{S}'_0$ (J/mol K)	$\hat{S}_{\text{theory}}$ (J/mol K)
120	0.0096	2922	155.1	25.5	167.0	166.7
150	0.0121	3693	162.7	25.2	171.7	171.5
200	0.0163	4961	172.5	25.3	177.8	177.7
250	0.0208	6243	180.1	26.0	182.6	182.5
300	0.0245	7578	186.4	27.5	186.2	186.2
350	0.0287	9006	192.1	29.7	189.4	189.4
400	0.0328	10,550	197.4	32.3	192.2	192.2

REFERENCE STATE (0): $T_0 = 298.15$ K, $p_0 = 0.101325$ MPa, $\hat{V}_0 = 0.0244$ m^3/mol, $\hat{S}_0 = 186.266$ J/mol K, $\hat{U}_0 = 7505$ J/mol

SOURCE: B. A. Younglove and J. F. Ely, *J. Phys. Chem. Ref. Data*, **16**, 577, 1987.

where

$$\hat{S}_{\text{ideal}} = R \ln \left[\frac{e^{5/2}(2\pi mkT)^{3/2}\hat{V}}{\tilde{N}h^3} \right] \tag{4.103}$$

In SI units, the constants are

$$R = 8.3136 \text{ J/(mol·K)}$$

$$m = 2.6578 \times 10^{-26} \text{ kg } (CH_4)$$

$$h = 6.62 \times 10^{-34} \text{ J·s}$$

$$k = 1.380 \times 10^{-23} \text{ J/(molecule·K)}$$

and

$$e = 2.71828$$

$$\tilde{N} = 6.02 \times 10^{23} \text{ molecules/mol}$$

giving the ideal gas contribution as

$$\hat{S}_{\text{ideal}} = 8.3136 \ \ln[T^{3/2}\hat{V}(2.4403 \times 10^5)] \tag{4.104}$$

Since in applications we are only interested in changes in $\hat{S}$ between two states, we can add an arbitrary reference state to our entropy equation as

$$\hat{S} = \hat{S}_0' + \left[\hat{S}_{\text{ideal}} - \frac{RT}{\hat{V}}\frac{dB}{dT} \right] \tag{4.105}$$

For consistency, we will use the same reference state values as those reported in Table 4.1; thus, from Eq. (4.105)

$$\hat{S}_0' = 186.266 \text{ J/mole} \cdot \text{K} - \left[\hat{S}_{\text{ideal}}(T_0, \hat{V}_0) - \frac{RT_0}{\hat{V}_0}\frac{dB}{dT}\bigg|_{T=T_0} \right]$$

$$= 42.82 - \frac{RT_0}{\hat{V}_0}\frac{dB}{dT}\bigg|_{T=T_0}$$

$$= 42.78 \text{ J/mol} \cdot \text{K}$$

The MATLAB program used to generate dB/dT is given in Table 4.2. The values of the entropy calculated from Eq. (4.105) are listed in the last column of Table 4.1. It is seen that corrections to the ideal gas entropy values are extremely small for this particular system.

4.6 Configurational Integral Equation for Dense Gases and Liquids

Now that we have learned how the thermodynamic properties are computed from knowledge of the configurational density functions for dilute gases, let's turn our attention to stronger interacting systems—dense gases and liquids. At high pressures and/or low temperatures, the

TABLE 4.2 MATLAB Program to Generate Temperature Derivative of the Second Virial Coefficient

```
% virderiv.m
% program to compute the temperature derivative of
% the second virial coefficient for a L-J (6-12) potential
%
% input parameters (SI units)
temp = 350
sigma = 3.817eϵ − 10
avo = 6.02e + 23
epsk = 148.2
tempd = temp/epsk
coeff = (2/3)*pi*avo*(sigma^3)/temp
%
% compute sum
term3 = −(1/4)*tempd^(−1/4)*[−(2^0.5)/4]*gamma(−1/4)
for n = 1 : 40;
term1 = [−(2*n + 1)/4]*tempd^[−(2*n + 1)/4]
term2 = [−(2^(n + 0.5)]/(4*gamma(n + 1)))*gamma ((2*n − 1)/4);
term3 = term3 + term1*term2;
end
dbdt' = coeff*term3
```

molecular number densities are sufficiently high so that more than two molecules interact at the same time. Determination of the configurational distribution in these many-bodied systems is indeed a formidable task, and we only give a brief glimpse at this important subject.

Let's look more carefully at the integral equation truncated at triplet interactions using the superposition approximation. Substituting Eq. (4.57) into Eq. (4.52) gives

$$kT\,\frac{\partial \ln n_2}{\partial \mathbf{r}_1} + \frac{\partial \Phi(\mathbf{r}_{12})}{\partial \mathbf{r}_1} = -\frac{1}{n^3}\int n_2(\mathbf{r}_{13})n_2(\mathbf{r}_{23})\frac{\partial \Phi(\mathbf{r}_{13})}{\partial \mathbf{r}_1}d\mathbf{r}_3 \tag{4.106}$$

or

$$kT\,\frac{\partial \ln n_2}{\partial \mathbf{r}_{12}} = -\frac{\partial \Phi(\mathbf{r}_{12})}{\partial \mathbf{r}_{12}} - \frac{1}{n^3}\int n_2(\mathbf{r}_{13})n_2(\mathbf{r}_{23})\frac{\partial \Phi(\mathbf{r}_{13})}{\partial \mathbf{r}_{13}}d\mathbf{r}_3 \tag{4.107}$$

Taking the dot product of the above equation with $(\mathbf{r}_{12}/r_{12})$ gives

$$kT\,\frac{\partial \ln n_2}{\partial r_{12}} = -\frac{\partial \Phi(\mathbf{r}_{12})}{\partial r_{12}} - \frac{1}{n^3}\int \frac{\partial \Phi(r_{13})}{\partial r_{13}}\left(\frac{\mathbf{r}_{12}}{r_{12}}\cdot\frac{\mathbf{r}_{13}}{r_{13}}\right)n_2(r_{13})n_2(r_{23})d\mathbf{r}_3 \tag{4.108}$$

Now, consider a spherical coordinate system with its center located on molecule (1) and radial coordinate r_{13} ($\mathbf{r}_{13} = \mathbf{r}_3 - \mathbf{r}_1$). Let $\mathbf{r}_{12}$ lie along the z-axis in this system. Thus

$$d\mathbf{r}_3 = d\mathbf{r}_{13} = r_{13}^2\,dr_{13}\sin\theta\,d\theta\,d\phi \tag{4.109}$$

In this coordinate system

$$\cos\theta = \frac{\mathbf{r}_{13}}{r_{13}} \cdot \frac{\mathbf{r}_{12}}{r_{12}} = \frac{r_{12}^2 + r_{13}^2 - r_{23}^2}{2r_{12}r_{13}} \tag{4.110}$$

and for a given azimuthal angle ϕ and r_{13}, Eq. 4.110 also yields

$$d(\cos\theta) = -\sin\theta d\theta = -\frac{r_{23}}{r_{12}r_{13}}dr_{23} \tag{4.111}$$

Substituting Eqs. (4.111) and (4.110) into Eq. (4.108), and carrying out the azimuthal angle integration, gives

$$kT\frac{\partial \ln n_2}{\partial r_{12}} = -\frac{\partial\Phi(\mathbf{r}_{12})}{\partial r_{12}} - \frac{\pi}{n^3}\int_0^\infty\int_{|r_{12}-r_{13}|}^{r_{12}+r_{13}}\frac{\partial\Phi}{\partial r_{13}} \times\left(\frac{r_{13}^2 + r_{12}^2 - r_{23}^2}{2r_{12}^2}\right)n_2(r_{13})n_2(r_{23})r_{23}\,dr_{23}\,dr_{13} \tag{4.112}$$

This last equation can be integrated to give

$$kT\ \ln g(r_{12}) = \int_{r_{12}}^\infty p(r_{12})dr_{12} \tag{4.113}$$

where

$$p(r_{12}) = \frac{d\,\Phi(r_{12})}{dr_{12}} + \pi n\int_0^\infty\int_{|r_{12}-r_{13}|}^{r_{12}+r_{13}}\frac{d\,\Phi(r_{13})}{dr_{13}} \times\left(\frac{r_{13}^2 + r_{12}^2 - r_{23}^2}{2r_{12}^2}\right)g(r_{23})g(r_{13})r_{23}\,dr_{23}\,dr_{13} \tag{4.114}$$

and $g(r_{12}) = n_2(r_{12})/n^2$

Now, since $\Phi(r_{12}) = 0$ at $r_{12} = \infty$, we obtain

$$kT\ \ln g(r_{12}) = -\Phi(r_{12}) + \pi n\int_{r_{12}}^\infty\frac{1}{r_{12}^2}\left\{\int_0^\infty\frac{d\,\Phi(r_{13})}{dr_{13}}g(r_{13}) \times\left(\int_{|r_{12}-r_{13}|}^{r_{12}+r_{13}}\left[(r_{13}^2 + r_{12}^2) - r_{23}^2\right]g(r_{23})r_{23}\,dr_{23}\right)dr_{13}\right\}dr_{12} \tag{4.115}$$

The above equation, for a given potential $\Phi(r_{ij})$, can be solved numerically by assuming a functional form of $g(r_{ij})$.† (Also see the third footnote of Sec. 4.3.) The result can be used to determine the thermodynamic properties U, S^G, and P from Eqs. (4.58), (4.68), and (4.45, 4.46),

†Some further analytical reductions of Eq. (4.115) are also possible. See Ref. 11.

respectively. Numerical solutions are beyond the scope of this chapter. Criticism and refinement of the superposition approximation is given by D. A. McQuarrie in the Further Reading section at the end of this chapter.

Interlude 4.2 Getting Microscopic Information from Macroscopic Observations: The Inverse Problem† When one thinks of equilibrium statistical mechanics, what usually comes to mind is its classical mission, namely, predicting macroscopic structure and properties from microscopic, pairwise interaction forces. However, more often than not, we are faced with a need to deduce information on the (conservative) pair interaction forces [equivalently, the pair potential $\Phi(r)$] from observed macroscopic properties. The information thus obtained can then be used to predict other properties of interest. The problem of extracting microscopic information from macroscopic observations is known as the *inverse problem*.

There are a number of "crude" methods which, when used with caution, lead to useful results.[3] An example is the use of experimentally measured second virial coefficient B to obtain the pair potential $\Phi(r)$ from the relation, Eq. (4.79),

$$B(T) = -2\pi \tilde{N} \int_0^{\infty} \{\exp[-\phi(r_{12})/kT] - 1\} r_{12}^2 \, dr_{12}$$

for an isotropic liquid phase. An approach requires suitable a priori assumptions concerning the functional form of the potential $\Phi(r)$.

However, a formal and more accurate method, which does not require any prior knowledge of the functional form of $\Phi(r)$, exists for extracting $\Phi(r)$ from the radial distribution function $g(r)$ (or, equivalently, from the static structure factor, which is related to $g(r)$ through a Fourier transformation).[12,13] The relation between $\Phi(r)$ and $g(r)$ is unique.[14] Therefore, the inverse problem is well-posed mathematically (although it is often ill-conditioned in practice since obtaining sufficiently accurate and complete experimental data on structure factors is very difficult). The inversion relies on the so-called Ornstein-Zernike equation, which represents a more sophisticated *closure method* as compared to the superposition approximation shown in this chapter (see the Further Reading section at the end of this chapter). A number of examples of inversion are discussed in references cited earlier. A sample result, for liquid copper near its melting point, extracted using neutron scattering data available in the literature, is shown in Fig. 4.4.[15] The pair potential thus obtained is practically identical to a piecewise reconstructed pair potential available in the literature and predicts diffusion coefficient and viscosity within experimental uncertainty.[16]

Solutions for inversion of this type represent an application of statistical mechanics that is not commonly seen in the literature and hold considerable potential for bridging the microscopic world to the macroscopic.

†Contributed by Raj Rajagopalan

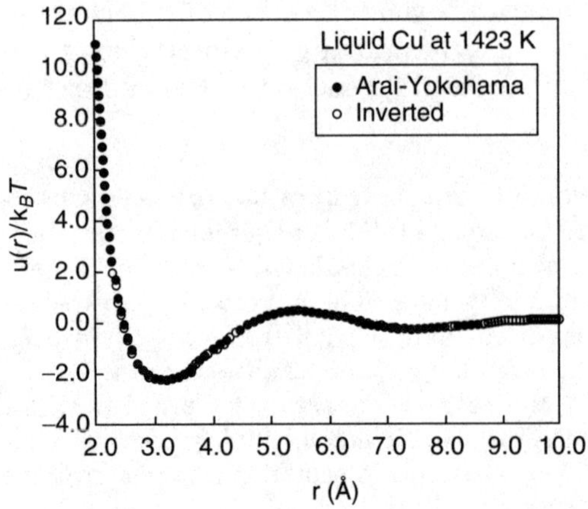

Figure 4.4 Inverted potential for liquid copper at T = 1423 K obtained from Ornstein-Zernike inversion by Rajagopalan and Srinivasa Rao.[14,15] Also shown is a piecewise reconstructed pair potential reported by Arai and Yokoyama.[16]

4.7 Equilibrium Properties of Nonspherical Molecules

The above descriptions of the equilibrium properties of matter were based on spherically symmetric intermolecular interactions. These descriptions work well for nonpolar, compact molecules as described in Chap. 1. Equilibrium properties for molecular systems described by nonspherical, asymmetric intermolecular interactions are considerably more difficult to treat algebraically due to the additional degrees of freedom. Nonetheless, we can follow the same procedures as in Secs. 4.1 and 4.2 to write the equilibrium density function as

$$f_N = \frac{\exp[-H(\mathbf{p}^N, \mathbf{q}^N)/kT]}{\frac{1}{N!}\int\int \exp[-H(\mathbf{p}^N, \mathbf{q}^N)/kT]d\,\mathbf{p}^N d\,\mathbf{q}^N} \tag{4.116}$$

where $(\mathbf{p}^N, \mathbf{q}^N)$ represents the entire set of generalized coordinates for the N-molecule system and $H(\mathbf{p}^N, \mathbf{q}^N)$ is the hamiltonian or total energy. The thermodynamic properties U and S also follow as, respectively,

$$U \equiv \frac{1}{N!}\int\int H(\mathbf{p}^N, \mathbf{q}^N)\, f_N\, d\,\mathbf{p}^N d\,\mathbf{q}^N \tag{4.117}$$

and

$$S \equiv -\frac{1}{N!}\frac{1}{h^{sN}}\int\int f_N' \ln f_N'\, d\,\mathbf{p}^N d\,\mathbf{q}^N \tag{4.118}$$

where s is the total number of generalized coordinates ($s = 5$ for a dipole) and $f'_N = h^{sN} f_N$. The equilibrium scalar pressure can be obtained from momentum conservation in generalized coordinates or from the partition function

$$Z_N = [N!h^{sN}]^{-1} \int\int \exp\{-H(\mathbf{p}^N, \mathbf{q}^N)/kT\} d\mathbf{p}^N d\mathbf{q}^N \tag{4.119}$$

(see Sec. 3.4 of Ref. 2).

Problem 4.11 involves the determination of U and S for a system of dipoles (dilute polar gas). The Further Reading section at the end of this chapter can be consulted for further treatments and examples.

This completes our description of the thermodynamic basis functions in terms of the configurational and momenta density functions obtained directly from the equilibrium solution to the Liouville equation. As will be shown in the next chapter, the nonequilibrium counterparts (local in space and time) of the thermodynamic basis functions can also be obtained directly from the Liouville equation, thus, providing a unified molecular view of equilibrium thermodynamics and chemical transport phenomena. Before moving on, however, we conclude this chapter by noting some important aspects of the equilibrium solution to the Liouville equation.

4.8 Complex Equilibrium Systems: Multicomponent and Multiphase

Some of the more intriguing problems in equilibrium statistical mechanics involve multicomponent and multiphase systems. For single phase, multicomponent systems we can readily extend the methodologies given here. For example, if N_α represents the number of molecules of component α, then

$$N = \sum_{\alpha=1}^{L} N_\alpha \tag{4.120}$$

where L is the total number of components present. Consider the expression for the internal energy, Eq. (4.36),

$$U = \frac{3}{2}NkT + \frac{1}{N!}\int n_N(\mathbf{r}^N)\Phi(\mathbf{r}^N)d\mathbf{r}^N$$

If we again assume pairwise additivity, we can contract Eq. (4.36) over any two molecules, where one molecule is with component $\alpha(\alpha \in 1, L)$

and the other with component $\beta(\beta \in 1, L)$ giving

$$U = \frac{3}{2}NkT + \frac{1}{2}V\sum_{\alpha=1}^{L}\sum_{\beta=1}^{L}\chi_\alpha\chi_\beta\int\phi(\mathbf{r}_{\alpha\beta})n_2(\mathbf{r}_{\alpha\beta})d\mathbf{r}_{\alpha\beta} \tag{4.121}$$

where $\mathbf{r}_{\alpha\beta}$ is the separation distance vector between a molecule of type α and a molecule of type β, and χ_α and χ_β are the number (or mole) fractions of species α and β, respectively.

Using the same type of arguments, it is possible to extend all thermodynamic functions in this manner. Of course, we have the added complexities of obtaining $n_2(\mathbf{r}_{\alpha\beta})$ in the multicomponent swarm of molecules, which is no easy task. A more extensive discussion can be found in the book by McQuarrie listed at the end of this chapter. (In particular, see pages 292–295.) Multiphase systems are even more complex, since more than one single equilibrium solution to the Liouville equation must exist. For more details, see the Further Reading section at the end of this chapter.

4.9 Chapter Summary and Note on the Lack of Proof of the Existence and Uniqueness of the Equilibrium Solution

Predicting the equilibrium macroscopic quantities of density, pressure, temperature, internal energy, and entropy has many practical applications in engineering and science. Not only are equilibrium states ubiquitous in practice, such as in holding and storage facilities, we also know, for example, that changes in internal energy from one equilibrium state to another equilibrium state are path independent and can be predicted from the equilibrium analysis given in this chapter. We have shown how having information on the intermolecular force interactions leads explicity to the determination of equilibrium thermodynamic properties. The global equilibrium analysis given in this chapter will also prove to be extremely useful in development of *local* equilibrium states in the treatment of nonequilibrium phenomena in Chaps. 5 and 6.

A word of caution is in order, however, to the development given here. In the preceding sections, we have obtained a particular solution to the steady-state Liouville equation, Eq. (4.1), under the equilibrium restrictions given by Eqs. (4.14) to (4.16). The question arises, however, as to whether or not any system beginning in some arbitrary initial state, if left to itself, would evolve to the time-independent equilibrium solution (existence question). The answer to this question is a definite "no" as there are many examples of interacting particle systems whose dynamics continually depend on its initial state. Furthermore, there may be other *steady-state solutions* to the Liouville equation, Eq. (4.1), satisfying Eqs. (4.14) to (4.16), besides the equilibrium solution (uniqueness issue). These questions are still the subject of current research and we refer the interested reader to the Further Reading section at the end of this chapter. However, it is still useful here to summarize a few of the key dynamical requirements on the system's capability of achieving an equilibrium state. First, the equations of motion describing the system dynamics must be *nonintegrable*, i.e., not capable of being integrated analytically. An example of an integrable system was given in Chap. 2 in solutions to the one-particle Liouville equation. Integrable systems cannot "forget" their initial state and therefore continuously depend on time. As we have seen here, one necessary requirement for an equilibrium state is that it is time independent. A second requirement is that the system be *ergodic*. In an ergodic system, by definition, any phase point will eventually sweep out all the available phase space (or come arbitrarily close to doing so) consistent with any constraints imposed on the system, such as the constant energy constraint of the microcanonical ensemble or constant temperature constraint of the canonical ensemble. The third and final requirement is that the system be *mixing*. By mixing it is meant that if we follow a volume element of phase space, the volume element does not retain its shape but will grossly distort over time. In mixing systems, the phase points are spreading out more uniformly over the available phase space. Ergodicity is a minimally or weakly necessary condition for an equilibrium distribution in that, by definition, in an equilibrium distribution every possible state (atomic positions and momenta) of the ensemble is equally likely. This condition is known as the *principle of equal a priori probabilities*, and it allows the derivation of the equilibrium distribution function from purely probability arguments.[†] Mixing leads to a more random distribution in phase point trajectories and is therefore a stronger condition for the establishment of an equilibrium state. An illustration of phase point trajectories showing the randomness and space-filling requirements for an equilibrium state in the canonical ensemble are shown in Fig. 4.5.

The systems capability to achieve an equilibrium state has important consequences in terms of so-called irreversible behavior. By *irreversible behavior* we mean a nonequilibrium system that demonstrates a specific direction or change in a property value, e.g., energy flux for irreversible systems is always in a direction opposite to the temperature gradient. Another example is the law of entropy increase. We will revisit these concepts in Chap. 6 when the transport property flux terms

[†]See the Further Reading section at the end of this chapter.

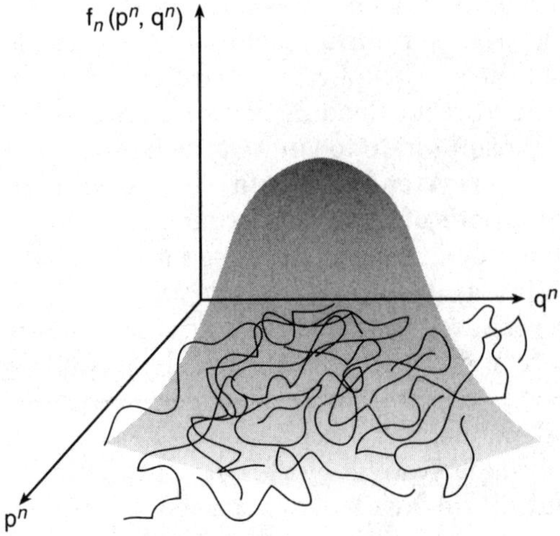

Figure 4.5 Illustration of phase point trajectories for an equilibrium distribution in the canonical ensemble. The trajectories sweep out the entire phase space randomly. Each trajectory corresponds to a particular total energy. The probability of finding a particular phase point follows the gaussian pattern shown. Note that for a microcanonical ensemble (not shown) the probability surface is uniform or flat and each phase point trajectory is at the same fixed total energy (cf. Fig. 4.1 and Fig. 4.2).

(mass, momentum, energy, and entropy) are considered in detail. But, the key feature of most of the results in transport phenomena and irreversible thermodynamics is that these nonequilibrium systems are generally assumed to be only slightly perturbed from local equilibrium behavior. The existence of local equilibrium states and perturbations thereof are often implied (sometimes incorrectly!) in the analysis of transport phenomena. As a counter example, systems of low dimensionality may not be capable of achieving such states (see the discussion and references on Poincare recurrence times in Chap. 3).

In the next chapter, we will consider the nonequilibrium behavior of matter in the most general way by deriving the spatial and temporal variations in density, average velocity, internal and kinetic energy, and entropy. We will use the formal definitions of these quantities introduced in this chapter, including the possibility of their spatial and temporal variations via the probability density function described by the full Liouville equation. In the next chapter, we will also formally define local equilibrium behavior and look at some specific, well-known examples of such behavior in science and engineering.

Problems

4.1 Utilizing the following integral

$$\int_{-\infty}^{\infty} \exp\left[-\frac{\beta x^2}{2m}\right] dx = \sqrt{\frac{2m\pi}{\beta}}$$

prove Eq. (4.30).

4.2 Noting that $n_N = W_N/Q_N$, prove Eqs. (4.37) and (4.42).

4.3 Referring to Ref. 1, prove Eq. (4.107).

4.4 By writing the dilute gas N-particle distribution function as

$$n_N = \frac{N!}{V^N}\frac{1}{n^2}[n_2(\mathbf{r}_{12}) + n_2(\mathbf{r}_{34}) + \cdots]$$

prove Eq. (4.115).

4.5 Beginning with the expression for the second virial coefficient

$$B(T) = +\frac{2\pi\tilde{N}}{3}\int_0^\infty \frac{d}{dr_{12}}\{\exp[-\phi(r_{12})/kT]\}r_{12}^3\,dr_{12}$$

prove, through integration by parts, that this is equivalent to

$$B(T) = -2\pi\tilde{N}\int_0^\infty \{\exp[-\phi(r_{12})/kT] - 1\}r_{12}^2\,dr_{12}$$

4.6 Prove the formula for the second virial coefficient for point centers of attraction or repulsion.

4.7 Derive an analytical expression for the second virial coefficient for the Lennard-Jones interaction potential. (See Ref. 2.)

4.8 The following experimental values of the second virial coefficient for argon in the pressure range 0–2900 atm have been reported [Michels et al., *Physica*, **15**, 627–633, 1949]

T (°C)	$B(T)$(cm^3/gmole)
0	−21.15
25	−18.47
50	−15.49
75	−11.62
100	−7.27
125	−2.30
150	+3.41

Determine the Lennard-Jones force constants for this system (see Ref. 2).

4.9 Let $\alpha = \alpha(p,q)$ be a constant of motion for the system, i.e.,

$$\frac{d\alpha}{dt} = \frac{\partial\alpha}{\partial p}\dot{p} + \frac{\partial\alpha}{\partial q}\dot{q} = 0$$

From the Liouville equation, in terms of the generalized coordinates (q) and conjugate momenta (p) (Chap. 2), show that a general condition for equilibrium is

$$\rho(p, q) = \rho(\alpha)$$

4.10 Prove Eq. (4.40) by carrying out the momentum integrations in Eq. (4.39) using the equilibrium solution, Eq. (4.34).

4.11 Derive expressions for $\hat{U}$ and $\hat{S}_G$ for the dilute polar gas (see Example 3.2).

4.12 Prove Eq. (4.49) by first integrating over all molecules except the set {h}, i.e.,

$$\int \frac{\partial n_N}{\partial \mathbf{r}_i} d\mathbf{r}^{N-h} = (N-h)! \frac{\partial n_h}{\partial \mathbf{r}_i} = \frac{1}{kT} \int n_N \frac{\partial \Phi(\mathbf{r}^n)}{\partial \mathbf{r}_i} d\mathbf{r}^{N-h}$$

where $i \in \{h\}$. Next write the potential in its pairwise additive form

$$\Phi(\mathbf{r}^N) = \sum_i \sum_{j<i} \Phi(\mathbf{r}_{ij})$$

and consider the following two cases separately: $j \in \{h\}$ and $j \notin \{h\}$

4.13 The Grand Canonical Ensemble: Consider the Liouville equation for a system in which the total molecule number N is a variable and no longer constant. By the same procedures used in Chap. 2, in this extended space we can readily derive the extended Liouville equation

$$\frac{\partial \rho_E}{\partial t} + \sum_{i=1}^{N} \frac{\partial \rho_E}{\partial q_i} \frac{\partial H}{\partial p_i} - \frac{\partial \rho_E}{\partial P_i} \frac{\partial H}{\partial q_i} + \frac{\partial \rho_E}{\partial N} = 0$$

where $\rho_E = \rho_E(\mathbf{q}^N, \mathbf{p}^N, N, t)$.

Now, following the same procedures as in Sec. 4.1, show by separation of variables that a particular steady-state solution to the extended Liouville equation can be written as

$$\rho_E(\mathbf{p}^N, \mathbf{q}^N, N) = [E^{\text{Grand}} N! h^{3N}]^{-1} \exp(\lambda N) \exp\left[-\frac{H(\mathbf{p}^N, \mathbf{q}^N)}{kT}\right]$$

where λ is a constant and E^{Grand} is called the grand canonical ensemble partition function given explicitly by

$$E^{\text{Grand}} = \sum_N E_N^{\text{Can}} e^{\lambda N}$$

where E_N^{Can} is the classical canonical partition function

$$E_N^{\text{Can}} = [N! h^{3N}]^{-1} \int\int \exp\left[-\frac{H(\mathbf{p}^N, \mathbf{q}^N)}{kT}\right] d\mathbf{p}^N d\mathbf{q}^N$$

Note that the constant λ can be related to the so-called chemical potential μ by

$$\lambda = \frac{\mu}{kT}$$

where $\mu = (\partial U/\partial N)_{S,V}$ (see the Further Reading section for more discussion on the grand canonical ensemble).

References

1. D. W. Heermann, *Computer Simulation Methods in Theoretical Physics*, 2nd ed., Springer-Verlag, New York, 1989. [An enlightening comparison of the canonical and microcanonical ensemble in molecular dynamics (computer experiments)].
2. C. F. Curtiss, J. O. Hirschfelder, and R. B. Bird, *Molecular Theory of Gases and Liquids*, Wiley, New York, 1960.
3. G. C. Maitland, M. Rigby, E. B. Smith, and W. A. Wakeham, *Intermolecular Forces—Their Origin and Determination*, Clarendon Press, Oxford, 1981.
4. M. Born and H. S. Green, A general kinetic theory of Liquids I. The molecular distribution functions, *Proc. Roy. Soc.*, London, **A188**, 10, 1946.
5. S. J. Yvon, *Actualities Scientifiques et Industrielles*, Hermann, Paris, 1935.
6. J. G. Kirkwood, V. A. Lewinson, and B. J. Alder, Radial distribution functions and the equations of state of fluids, *J. Chem. Phys.*, **20**, 929, 1952.
7. J. G. Kirkwood, Radial distribution functions in liquids, *J. Chem. Phys.*, **10**, 394, 1942.
8. B. B. Laird and A. D. J. Haymet, Calculation of the entropy of binary hard sphere mixtures from pair correlation functions. *J. Chem. Phys.*, **97**, 2153, 1992.
9. H. S. Green, *Molecular Theory of Fluids*, North-Holland, Amsterdam, 1952.
10. R. E. Nettleton and H. S. Green, Expression in terms of molecular distribution functions for the entropy density in an infinite system, *J. Chem. Phys.*, **29**, 1365, 1958.
11. T. L. Hill, *Statistical Mechanics*, McGraw-Hill, New York, 1956, p. 207.
12. R. Rajagopalan and K. Srinivasa Rao, Interaction forces in charged colloids: Inversion of static structure factors, *Phys. Rev. E*, **55**, 4423, 1997 [and the references cited therein].
13. K. Srinivasa Rao and R. Rajagopalan, Interaction forces in quasi two dimensional charged dispersions, *Phys. Rev. E*, **57**, 3227, 1998 [and the references cited therein].
14. J. R. Henderson, On the test particle approach to the statistical mechanics of hard-sphere fluids, *Mol. Phys.*, **48**, 389, 1983.
15. Y. Waseda, *The Structure of Non-Crystalline Materials*, McGraw-Hill, New York, 1980.
16. T. Arai and I. Yokoyama, Effecting pair potential reproducing the measured structure factor of liquid Cu near the melting point, *J. Phys. Condens. Matter*, **3**, 7475, 1991.

Further Reading

C. Kittel, *Statistical Physics*, Wiley, New York, 1958. [An excellent introductory discussion of the various types of ensembles is given in this easy to read book.]

D. A. McQuarrie, *Statistical Mechanics*, Harper & Row, New York, 1976. [This well-known book provides an extensive treatment of the statistical thermodynamics of gases, liquids, and solids. Chapter 13 provides a comprehensive description of configurational integral equations. Also see the discussion in J. M. Ziman, *Models of Disorder: The Theoretical Physics of Homogeneously Disordered Systems,* Cambridge University Press, Cambridge 1979.]

I. Prigogine, *From Being to Becoming, Time and Complexity in the Physical Sciences,* W. H. Freeman, New York, 1980. [An excellent introductory treatment on aspects of existence and uniqueness of the equilibrium solution. See also I. Prigogine and I. Stengers, *Order Out of Chaos, Man's New Dialogue with Nature*, Bantam Books, New York, 1984. Ilya Prigogine received the 1977 Nobel Prize in Chemistry for his contributions to nonequilibrium thermodynamics.]

R. C. Tolman, *The Principles of Statistical Mechanics*, Dover, New York, 1979. [More advanced discussion on ensemble types is provided by this classic text in statistical mechanics. Also, see Hill.[11]]

T. Boublik, I. Nezbeda, and K. Hlavaty, *Statistical Thermodynamics of Simple Liquids and Their Mixtures*, Elsevier, New York, 1980 [For some introductory references on multicomponent and multiphase systems, see this reference and also see H. E. Stanley, *Introduction to Phase Transitions and Critical Phenomena*, Oxford University Press, New York, 1971.]

W. G. Hoover, *Computational Statistical Mechanics*, Elsevier, New York, 1991. [An alternative derivation of the canonical distribution is possible by incorporating temperature as a constraint in the equations of motion of the molecules. The resulting constrained equations are known as Nosé-Hoover mechanics. For more details, see this reference and the references cited therein.]

Chapter

5

The General Equations of Change for Nonequilibrium Systems

"Welcome to sweet, sunny Solla Sollew
Where we never have troubles
At least very few.
As a matter of fact, we have only just one.
Imagine! Just one little trouble, my son."

—THEODOR SEUSS GEISEL,
I Had Trouble in Getting to Solla Sollew

5.1 Introduction: The Scope of Transport Phenomena

In the previous chapter, we studied equilibrium expressions for the thermodynamic basis functions $\hat{V}$, $\mathbf{P}$, $\hat{U}$, and $\hat{S}$. For equilibrium systems, these functions are spatially and temporally independent. In nonequilibrium systems, on the other hand, these functions can depend on both space and time. Furthermore, as will be shown in this chapter, their nonequilibrium behavior is described by the so-called *transport equations* or *conservation equations* that can be obtained directly from the Liouville equation. Specifically, we have the following relationships that will be established:

Nonequilibrium basis functions		Corresponding conservation equation
$\hat{V}(\mathbf{r}, t)$	$\longrightarrow$	mass conservation
$\mathbf{P}(\mathbf{r}, t)$	$\longrightarrow$	momentum conservation
$\hat{U}(\mathbf{r}, t)$	$\longrightarrow$	energy conservation
$\hat{S}(\mathbf{r}, t)$	$\longrightarrow$	entropy conservation

The Liouville equation, Eq. (2.6), describes the behavior of the collection of phase points as they move through a multidimensional space, or phase space, representing the position and momentum coordinates of all molecules in the system. The phase points tend to be concentrated in regions of phase space where it is most likely to find the N molecules with a certain momentum and position. Thus, the density function ρ_N can be interpreted (aside from a normalization constant) as a probability density function, i.e., $\rho_N d\mathbf{r}^N d\mathbf{p}^N$ is proportional to the probability of finding a phase point in a multidimensional region between $(\mathbf{r}^N, \mathbf{p}^N)$ and $(\mathbf{r}^N + d\mathbf{r}^N, \mathbf{p}^N + d\mathbf{p}^N)$ at any time t.

Just as one defines the mean, variance, and other moments of probability density functions, we can also examine these quantities with respect to the density function ρ_N. More specifically, as shown originally by Irving and Kirkwood,[1] this averaging can be performed directly with the Liouville equation leading to the so-called transport equations. Since the Liouville equation is a conservation equation, the transport equations also represent conservation equations for the various moments of the density function ρ_N. The moments will be defined more specifically in this chapter.

To preview the results somewhat, it will be shown that the general form of the transport equations contains expressions for the property flux variables (momentum flux $\mathbf{P}$, energy flux $\mathbf{q}$, and entropy flux $\mathbf{s}$) involving integrals over lower-order density functions. In this form, the transport equations are referred to as *general equations of change* since virtually no assumptions are made in their derivation. In order to finally resolve the transport equations, expressions for specific lower-ordered distribution functions must be determined. These are, in turn, obtained from solutions to reduced forms of the Liouville equation, and this is where critical approximations are usually made. For example, the Euler and Navier-Stokes equations of motion derived in the next chapter have flux expressions based on certain approximate solutions to reduced forms of the Liouville equation. Let's first look, however, at the most general forms of the transport equations.

5.2 Conservation Equation for Any Dynamical Variable

Some commonalities exist in the derivation of the mass, momentum, and energy conservation equations from the Liouville equation, and thus, the development of these equations can be expedited by first considering an arbitrary quantity $\alpha(\mathbf{r}^N, \mathbf{p}^N)$ that does not depend explicitly on time.

The average or expectation value of α is defined as

$$\langle \alpha \rangle = \frac{1}{N!} \int_{-\infty}^{+\infty} \int_{-\infty}^{+\infty} \alpha(\mathbf{r}^N, \mathbf{p}^N) f_N(\mathbf{r}^N, \mathbf{p}^N, t)\, d\mathbf{r}^N\, d\mathbf{p}^N \qquad (5.1)$$

Multiplying the Liouville equation, Eq. (2.6), written in terms of f_N, by $\alpha(\mathbf{r}^N, \mathbf{p}^N)/N!$ and integrating over all phase space gives

$$\frac{1}{N!}\int\int \alpha \frac{\partial f_N}{\partial t} d\mathbf{r}^N \, d\mathbf{p}^N$$
$$= -\frac{1}{N!}\frac{1}{m}\int\int \alpha \left(\sum_{i=1}^{N} \mathbf{p}_i \cdot \frac{\partial f_N}{\partial \mathbf{r}_i}\right) d\mathbf{r}^N d\mathbf{p}^N$$
$$+\frac{1}{N!}\int\int \alpha \left(\sum_{i=1}^{N} \frac{\partial \Phi}{\partial \mathbf{r}_i} \cdot \frac{\partial f_N}{\partial \mathbf{p}_i}\right) d\mathbf{r}^N d\mathbf{p}^N \tag{5.2}$$

where the integrations extend all over the phase space. Since α, as assumed, does not depend explicitly on time, the term on the left-hand side of Eq. (5.2) can be written as

$$\frac{1}{N!}\int\int \alpha \frac{\partial f_N}{\partial t} d\mathbf{r}^N \, d\mathbf{p}^N = \frac{\partial}{\partial t}\left[\frac{1}{N!}\int\int \alpha f_N d\mathbf{r}^N d\mathbf{p}^N\right] = \frac{\partial \langle \alpha \rangle}{\partial t}. \tag{5.3}$$

Using the vector identity given generically by

$$a\mathbf{b} \cdot \frac{\partial c}{\partial \mathbf{r}} = \frac{\partial}{\partial \mathbf{r}} \cdot [ac\mathbf{b}] - c\frac{\partial}{\partial \mathbf{r}} \cdot (a\mathbf{b}) \tag{5.4}$$

where $\mathbf{b}$ is an arbitrary vector and $\partial/\partial\mathbf{r}$ is a del operator, the first term on the right-hand side of Eq. (5.2) can be expressed as

$$-\frac{1}{N!}\int\int \alpha \frac{1}{m}\left(\sum_{i=1}^{N} \mathbf{p}_i \cdot \frac{\partial f_N}{\partial \mathbf{r}_i}\right) d\mathbf{r}^N d\mathbf{p}^N$$
$$= \underline{-\frac{1}{N!}\int\int \frac{1}{m}\left(\sum_{i=1}^{N} \frac{\partial}{\partial \mathbf{r}_i} \cdot [\alpha \mathbf{p}_i f_N] d\mathbf{r}^N\right) d\mathbf{p}^N}$$
$$+\frac{1}{N!}\int\int \frac{1}{m}\left[\sum_{i=1}^{N} \mathbf{p}_i f_N \frac{\partial \alpha}{\partial \mathbf{r}_i}\right] \tag{5.5}$$

The dashed underlined term vanishes by the application of Gauss' theorem, Eq. (3.3), which is shown as follows:

Writing the dashed underlined term in Eq. (5.5) as

$$-\frac{1}{N!}\int\int \frac{1}{m}\left(\sum_{i=1}^{N} \frac{\partial}{\partial \mathbf{r}_i} \cdot [\alpha \mathbf{p}_i f_N]\right) d\mathbf{r}^N d\mathbf{p}^N$$
$$= -\frac{1}{N!}\left(\frac{1}{m}\right)\sum_{i=1}^{N}\int\left\{\int \frac{\partial}{\partial \mathbf{r}_i} \cdot [\alpha \mathbf{p}_i f_N] d\mathbf{r}^N\right\} d\mathbf{p}^N \tag{5.6}$$

and using Gauss' theorem, the term in curly brackets in the above equation can be expressed as

$$\int_{-\infty}^{+\infty} \frac{\partial}{\partial \mathbf{r}_i} \cdot [\alpha\, \mathbf{p}_i f_N] d\mathbf{r}^N = \int \left[\int_{S_i \to \infty} \alpha \mathbf{p}_i f_N \cdot \mathbf{n}_i d S_i \right] d\mathbf{r}^{N-i} \qquad (5.7)$$

where $d\mathbf{r}^{N-i}$ is shorthand notation for $d\mathbf{r}_1, \ldots d\mathbf{r}_j, \ldots d\mathbf{r}_N, j \neq i$. Using the properties of f_N, Eq. (3.17), discussed previously, the surface integral term in Eq. (5.7) is identically zero and the dashed underlined term in Eq. (5.5) vanishes (QED).

By identical arguments it can be shown that the second term on the right-hand side of Eq. (5.2) can be expressed as

$$\frac{1}{N!} \int \int \alpha \left(\sum_{i=1}^{N} \frac{\partial \Phi}{\partial \mathbf{r}_i} \cdot \frac{\partial f_N}{\partial \mathbf{p}_i} \right) d\mathbf{r}^N d\mathbf{p}^N$$

$$= -\frac{1}{N!} \int \int \left(\sum_{i=1}^{N} f_N \frac{\partial \Phi}{\partial \mathbf{r}_i} \cdot \frac{\partial \alpha}{\partial \mathbf{p}_i} \right) d\mathbf{r}^N d\mathbf{p}^N \qquad (5.8)$$

The conservation equation for α, Eq. (5.2), finally becomes

$$\frac{\partial \langle \alpha \rangle}{\partial t} = \frac{1}{N!} \left(\frac{1}{m} \right) \int \int \sum_{i=1}^{N} \left(f_N \mathbf{p}_i \cdot \frac{\partial \alpha}{\partial \mathbf{r}_i} \right) d\mathbf{r}^N d\mathbf{p}^N$$

$$- \frac{1}{N!} \int \int \left(\sum_{i=1}^{N} f_N \frac{\partial \Phi}{\partial \mathbf{r}_i} \cdot \frac{\partial \alpha}{\partial \mathbf{p}_i} \right) d\mathbf{r}^N d\mathbf{p}^N \qquad (5.9)$$

5.3 Mass Conservation Equation (the Equation of Continuity)

The equation of mass conservation is obtained from Eq. (5.9) by letting

$$\alpha = \sum_{k=1}^{N} \delta(\mathbf{r}_k - \mathbf{r}) \qquad (5.10)$$

where $\mathbf{r}$ is a vector that locates an arbitrary point in the system (system locator vector), thus, the expectation value of α, $\langle \alpha \rangle$, is

$$\langle \alpha \rangle = \frac{1}{N!} \int \int \sum_{k=1}^{N} \delta(\mathbf{r}_k - \mathbf{r}) f_N(\mathbf{r}^N, \mathbf{p}^N, t) d\mathbf{r}^N d\mathbf{p}^N \qquad (5.11)$$

Using the reduced density function definitions from Chap. 3

$$\langle \alpha \rangle = \frac{1}{N!} \sum_{k=1}^{N} (N-1)! \int\int \delta(\mathbf{r}_k - \mathbf{r}) f_1(\mathbf{r}_k, \mathbf{p}_k, t) d\mathbf{r}_k d\mathbf{p}_k$$

$$= \frac{1}{N} \sum_{k=1}^{N} \int \delta(\mathbf{r}_k - \mathbf{r}) n(\mathbf{r}_k, t) d\mathbf{r}_k = n(\mathbf{r}, t) \tag{5.12}$$

which is the average or macroscopic number density at location $\mathbf{r}$ and at time t. Recall that $n(\mathbf{r}_k, t)\, d\mathbf{r}_k$ is, aside from a multiplicative constant, the probability of finding *any one* molecule between $\mathbf{r}_k$ and $\mathbf{r}_k + d\mathbf{r}_k$ at time t. Likewise, $n(\mathbf{r}, t)\, d\mathbf{r}$ is, aside from a multiplicative constant, the probability of finding *any one* molecule between $\mathbf{r}$ and $\mathbf{r} + d\mathbf{r}$ at time t; this is the probability interpretation of the ordinary macroscopic number density.

Now analyzing the first term on the right-hand side of Eq. (5.9) as

$$\frac{1}{N!}\left(\frac{1}{m}\right) \int\int \sum_{i=1}^{N} \left[f_N \mathbf{p}_i \cdot \frac{\partial}{\partial \mathbf{r}_i} \left(\sum_{k=1}^{N} \delta(\mathbf{r}_k - \mathbf{r}) \right) \right] d\mathbf{r}^N d\mathbf{p}^N$$

$$= \frac{1}{N!}\left(\frac{1}{m}\right) \sum_{i=1}^{N} \int\int f_N \mathbf{p}_i \cdot \left[\frac{\partial}{\partial \mathbf{r}_i} \delta(\mathbf{r}_i - \mathbf{r}) \right] d\mathbf{r}^N d\mathbf{p}^N \tag{5.13}$$

Introducing the reduced density function f_1 the above term becomes

$$\frac{(N-1)!}{N!}\left(\frac{1}{m}\right) \sum_{i=1}^{N} \int\int f_1(\mathbf{r}_i, \mathbf{p}_i, t) \mathbf{p}_i \cdot \left[\frac{\partial}{\partial \mathbf{r}_i} \delta(\mathbf{r}_i - \mathbf{r}) \right] d\mathbf{r}_i d\mathbf{p}_i$$

$$= \frac{1}{N} \sum_{i=1}^{N} \int n\mathbf{v}_0(\mathbf{r}_i, t) \cdot \left[\frac{\partial}{\partial \mathbf{r}_i} \delta(\mathbf{r}_i - \mathbf{r}) \right] d\mathbf{r}_i \tag{5.14}$$

where the *mass average velocity* $\mathbf{v}_0(\mathbf{r}, t)$ is defined as

$$\mathbf{v}_0(\mathbf{r}, t) \equiv \frac{1}{n} \int \left(\frac{\mathbf{p}}{m}\right) f_1(\mathbf{r}, \mathbf{p}, t)\, d\mathbf{p} \tag{5.15}$$

Finally, using the relationship for integrals involving the derivative of the delta function[2]

$$\int g(x)\delta^{(n)}(x - x_0) dx = (-1)^n g^{(n)}(x_0) \tag{5.16}$$

where $\delta^{(n)}$ indicates the nth derivative of δ with respect to x and, similarly, $g^{(n)}(x_0)$ is the nth derivative of g with respect to x evaluated at x_0, we obtain

$$\frac{1}{N}\sum_{i=1}^{N}\int n\mathbf{v}_0(\mathbf{r}_i,t)\cdot\left[\frac{\partial}{\partial\mathbf{r}_i}\delta\left(\mathbf{r}_i-\mathbf{r}\right)\right]d\,\mathbf{r}_i=-\frac{\partial}{\partial\mathbf{r}}\cdot\left[n\mathbf{v}_0(\mathbf{r},t)\right] \tag{5.17}$$

Also, the remaining term on the right-hand side of Eq. (5.9) is zero since α in this case is not a function of the momentum coordinates.

Finally, from Eqs. (5.12) and (5.17), the equation for mass conservation is obtained as

$$\frac{\partial n}{\partial t}+\frac{\partial}{\partial\mathbf{r}}\cdot\left[n\mathbf{v}_0\right]=0 \tag{5.18}$$

or, since we have assumed constant molecular mass

$$\frac{\partial\rho}{\partial t}+\frac{\partial}{\partial\mathbf{r}}\cdot\left[\rho\mathbf{v}_0\right]=0 \tag{5.19}$$

where $\rho\equiv mn$ is the ordinary mass density of the fluid.

Of course, for incompressible fluids ρ is a constant and we obtain the familiar result

$$\frac{\partial}{\partial\mathbf{r}}\cdot\mathbf{v}_0(\mathbf{r},t)=0\qquad\text{(incompressible fluids)} \tag{5.20}$$

i.e., the velocity field is divergentless for incompressible fluids.

Interlude 5.1 More About the Delta Function The delta function utilized extensively in this chapter is part of a group of similar functions known as generalized functions. These functions are defined, in part, in terms of their integral properties. For example, the delta function is defined by the following properties:

1.

$$\int_{\text{all }x}\delta\left(x-x_0\right)dx=1$$

2.

$$\delta\left(x-x_0\right)=\begin{cases}0 & x\neq x_0\\ \infty & x=x_0\end{cases}$$

There are numerous types of real functions that in some defined limit satisfy the above conditions, such as

$$\delta\left(x\right)=\lim_{\sigma\to 0}\frac{1}{\sqrt{\pi}\sigma}e^{-x^2/\sigma^2}$$

$$\delta\left(x\right)=\lim_{k\to\infty}\begin{cases}k & 0<x<1/k\\ 0 & x\leq 0\quad\text{and}\quad x\geq 1/k\end{cases}$$

and many others. These functions can be utilized to establish other properties of the delta function. Consider, for example,

$$\int_{\text{all } x} g(x)\delta(x - x_0)dx$$

where g is an arbitrary real function. Using the limiting function approach

$$\begin{aligned}\int_{\text{all } x} g(x)\delta(x - x_0)dx &= \lim_{k\to\infty} k \int_0^{1/k} g(x')dx' \\ &= \lim_{k\to\infty} k \cdot g(\xi') \cdot \frac{1}{k}\end{aligned}$$

where $x' \equiv x - x_0$ and the last step follows from the mean value theorem of integral calculus. Thus

$$\int_{\text{all } x} g(x)\delta(x - x_0)dx = \lim_{k\to\infty} g(\xi') = g(x_0)$$

where we note that $\xi' \to 0$ as $k \to \infty$, or $x \to x_0$ as $k \to \infty$.

As a somewhat more complicated example, consider the derivative operation

$$\int_{\text{all } x} g(x)\delta'(x - x_0)dx$$

where $\delta'(x - x_0) \equiv d[\delta(x - x_0)]/dx$. Integrating by parts

$$\begin{aligned}&\int_{\text{all } x} g(x)\delta'(x - x_0)dx \\ &\quad = \delta(x - x_0)g(x)|_{-\infty}^{\infty} - \int_{\text{all } x} \delta(x - x_0)g'(x)dx\end{aligned}$$

The first term on the right-hand side vanishes unless $g(x)$ is infinite at the boundaries $x = \pm\infty$. The second term is simply $g'(x_0)$, thus

$$\int g(x)\delta'(x - x_0)dx = -g'(x_0)$$

Now, try to prove the corresponding integral formula for the nth derivative $\delta^{(n)}(x - x_0)$.

5.4 Momentum Conservation

The momentum conservation equation is obtained by setting

$$\alpha = \sum_{k=1}^{N} \mathbf{p}_k \delta(\mathbf{r}_k - \mathbf{r}) \tag{5.21}$$

where again $\mathbf{r}$ is a system locator vector. Thus, the expectation value of α, $\langle\alpha\rangle$, becomes

$$\begin{aligned}\langle\alpha\rangle &= \frac{1}{N!}\int\int\sum_{k=1}^{N}\mathbf{p}_k\delta(\mathbf{r}_k-\mathbf{r})f_N(\mathbf{r}^N,\mathbf{p}^N,t)\,d\mathbf{r}^N d\mathbf{p}^N \\ &= \frac{1}{N}\sum_{k=1}^{N}\int\int\mathbf{p}_k\delta(\mathbf{r}_k-\mathbf{r})f_1(\mathbf{r}_k,\mathbf{p}_k,t)\,d\mathbf{r}_k d\mathbf{p}_k \\ &= \frac{1}{N}\sum_{k=1}^{N}\int nm\mathbf{v}_0(\mathbf{r}_k,t)\delta(\mathbf{r}_k-\mathbf{r})\,d\mathbf{r}_k = \rho\mathbf{v}_0(\mathbf{r},t)\end{aligned} \qquad (5.22)$$

where $\rho\mathbf{v}_0$ is the macroscopic momentum density (momentum/volume).

The first term on the right-hand side of Eq. (5.9) becomes

$$\begin{aligned}&\frac{1}{N!}\left(\frac{1}{m}\right)\sum_{i=1}^{N}\int\int f_N\mathbf{p}_i\cdot\sum_{k=1}^{N}\mathbf{p}_k\left[\frac{\partial}{\partial\mathbf{r}_i}\delta(\mathbf{r}_k-\mathbf{r})\right]d\mathbf{r}^N d\mathbf{p}^N \\ &= \frac{1}{N}\left(\frac{1}{m}\right)\sum_{i=1}^{N}\int\int f_1(\mathbf{r}_i,\mathbf{p}_i,t)\mathbf{p}_i\cdot\left[\mathbf{p}_i\frac{\partial}{\partial\mathbf{r}_i}\delta(\mathbf{r}_i-\mathbf{r})\right]d\mathbf{r}_i d\mathbf{p}_i\end{aligned} \qquad (5.23)$$

Now, rearranging the vector-dyadic product term in the last equation as

$$\mathbf{p}_i\cdot\left[\mathbf{p}_i\frac{\partial}{\partial\mathbf{r}_i}\delta(\mathbf{r}_i-\mathbf{r})\right] = (\mathbf{p}_i\mathbf{p}_i)\cdot\frac{\partial}{\partial\mathbf{r}_i}\delta(\mathbf{r}_i-\mathbf{r}) \qquad (5.24)$$

and using Eqs. (5.16) and (5.23) becomes

$$\frac{1}{N}\left(\frac{1}{m}\right)\frac{\partial}{\partial\mathbf{r}}\cdot\sum_{i=1}^{N}\int\mathbf{p}_i\mathbf{p}_i f_1(\mathbf{r},\mathbf{p}_i,t)\,d\mathbf{p}_i = \frac{1}{m}\frac{\partial}{\partial\mathbf{r}}\cdot\int\mathbf{p}\mathbf{p}f_1(\mathbf{r},\mathbf{p},t)\,d\mathbf{p} \qquad (5.25)$$

where $\mathbf{p}$ is the momentum of any one molecule. The above term can be alternatively expressed relative to the mass average velocity, Eq. (5.15), as (Prob. 5.1)

$$\frac{1}{m}\frac{\partial}{\partial\mathbf{r}}\cdot\int\mathbf{p}\mathbf{p}f_1(\mathbf{r},\mathbf{p},t)\,d\mathbf{p} = \frac{\partial}{\partial\mathbf{r}}\cdot[\rho\mathbf{v}_0\mathbf{v}_0+\mathbf{P}_k] \qquad (5.26)$$

where $\mathbf{P}_k$ is the so-called kinetic contribution to the pressure tensor, defined as

$$\mathbf{P}_k \equiv m\int\left(\frac{\mathbf{p}}{m}-\mathbf{v}_0\right)\left(\frac{\mathbf{p}}{m}-\mathbf{v}_0\right)f_1(\mathbf{r},\mathbf{p},t)\,d\mathbf{p} \qquad (5.27)$$

Finally, the second term on the right-hand side of Eq. (5.9) is, in this case,

$$-\frac{1}{N!}\int\int\sum_{i=1}^{N}\left[f_N\frac{\partial\Phi}{\partial\mathbf{r}_i}\cdot\frac{\partial}{\partial\mathbf{p}_i}\sum_{k=1}^{N}\mathbf{p}_k\delta\left(\mathbf{r}_k-\mathbf{r}\right)\right]d\,\mathbf{r}^N d\,\mathbf{p}^N$$

$$=-\frac{1}{N!}\sum_{i=1}^{N}\int\int f_N\frac{\partial\Phi}{\partial\mathbf{r}_i}\delta\left(\mathbf{r}_i-\mathbf{r}\right)d\,\mathbf{r}^N d\,\mathbf{p}^N \tag{5.28}$$

where we have used the vector identities

$$\frac{\partial}{\partial\mathbf{r}}\mathbf{r}=\mathbf{I} \tag{5.29}$$

and

$$\mathbf{b}\cdot\mathbf{I}=\mathbf{b} \tag{5.30}$$

In this last equation $\mathbf{b}$ is any vector and $\mathbf{I}$ is a *unit tensor*, which in matrix notation is given by

$$\mathbf{I}=\begin{bmatrix}1&0&0\\0&1&0\\0&0&1\end{bmatrix} \tag{5.31}$$

Integrating Eq. (5.28) over all momentum space we obtain

$$-\frac{1}{N!}\sum_{i=1}^{N}\int n_N\frac{\partial\Phi}{\partial\mathbf{r}_i}\delta\left(\mathbf{r}_i-\mathbf{r}\right)d\,\mathbf{r}^N \tag{5.32}$$

The potential Φ consists of an intermolecular potential and a potential due to an external force, as

$$\frac{\partial\Phi}{\partial\mathbf{r}_i}=\frac{\partial\Phi_{\text{int}}(\mathbf{r}^N)}{\partial\mathbf{r}_i}+\frac{\partial\Phi_{\text{ext}}(\mathbf{r}_i)}{\partial\mathbf{r}_i} \tag{5.33}$$

Thus, Eq. (5.32) becomes

$$-\frac{1}{N!}\sum_{i=1}^{N}\int n_N\frac{\partial\Phi_{\text{int}}(\mathbf{r}^N)}{\partial\mathbf{r}_i}\delta\left(\mathbf{r}_i-\mathbf{r}\right)d\,\mathbf{r}^N$$

$$-\frac{1}{N!}\sum_{i=1}^{N}\int n_N\frac{\partial\Phi_{\text{ext}}(\mathbf{r}_i)}{\partial\mathbf{r}_i}\delta\left(\mathbf{r}_i-\mathbf{r}\right)d\,\mathbf{r}^N \tag{5.34}$$

Integrating over the position space of the ith molecule, the second term in Eq. (5.34) becomes

$$-\frac{1}{N!}\sum_{i=1}^{N}\int n_N\frac{\partial\Phi_{\text{ext}}(\mathbf{r}_i)}{\partial\mathbf{r}_i}\delta\left(\mathbf{r}_i-\mathbf{r}\right)d\,\mathbf{r}^N=-n(\mathbf{r},t)\frac{\partial\Phi_{\text{ext}}(\mathbf{r})}{\partial\mathbf{r}} \tag{5.35}$$

Now, we will assume pairwise additivity of the interaction force (cf. Chap. 4, Sec. 4) as

$$\frac{\partial \Phi_{\text{int}}(\mathbf{r}^N)}{\partial \mathbf{r}_i} = \sum_{\substack{k=1 \\ k \neq i}}^{N} \frac{\partial \Phi_{\text{int}}(\mathbf{r}_{ik})}{\partial \mathbf{r}_i} \tag{5.36}$$

where $\mathbf{r}_{ik} = \mathbf{r}_k - \mathbf{r}_i$. Substituting into the first term of Eq. (5.34) and integrating gives

$$\begin{aligned} & -\frac{1}{N!}\sum_{i=1}^{N}\int n_N \frac{\partial \Phi_{\text{int}}(\mathbf{r}^N)}{\partial \mathbf{r}_i}\delta(\mathbf{r}_i - \mathbf{r})\, d\mathbf{r}^N \\ & = -\frac{1}{N!}\sum_{i=1}^{N}\sum_{\substack{k=1 \\ k \neq i}}^{N}(N-2)! \int n_2(\mathbf{r}_i, \mathbf{r}_k, t)\frac{\partial \Phi_{\text{int}}(\mathbf{r}_{ik})}{\partial \mathbf{r}_i}\delta(\mathbf{r}_i - \mathbf{r})\, d\mathbf{r}_i d\mathbf{r}_k \\ & = \int n_2(\mathbf{r}, \mathbf{r}', t)\frac{\partial \Phi_{\text{int}}(\mathbf{R})}{\partial \mathbf{R}}\, d\mathbf{R} \end{aligned} \tag{5.37}$$

where $\mathbf{R} = \mathbf{r}' - \mathbf{r}$, and we have used the relationship

$$\Phi_{\text{int}}(\mathbf{R}) = \Phi_{\text{int}}(-\mathbf{R}) \tag{5.38}$$

Next, we transform to center of mass and relative coordinates defined, respectively, by

$$\mathbf{r}_{\text{cm}} = \frac{1}{2}(\mathbf{r} + \mathbf{r}') \tag{5.39}$$

$$\mathbf{R} = \mathbf{r}' - \mathbf{r} \tag{5.40}$$

Thus,

$$\mathbf{r}_{\text{cm}} = \mathbf{r} + \frac{1}{2}\mathbf{R} \tag{5.41}$$

and

$$n_2(\mathbf{r}, \mathbf{r}', t) \equiv \tilde{n}_2(\mathbf{r}_{\text{cm}}, \mathbf{R}, t) \equiv \tilde{n}_2\left(\mathbf{r} + \frac{1}{2}\mathbf{R}, \mathbf{R}, t\right) \tag{5.42}$$

Now, expanding the center of mass about $\mathbf{r}$ gives

$$\tilde{n}_2(\mathbf{r} + \frac{1}{2}\mathbf{R}, \mathbf{R}, t) = \tilde{n}_2(\mathbf{r}, \mathbf{R}, t) + \frac{1}{2}\mathbf{R}\cdot\frac{\partial \tilde{n}_2(\mathbf{r}, \mathbf{R}, t)}{\partial \mathbf{r}} + \cdots \tag{5.43}$$

Note that the function $\tilde{n}_2(\mathbf{r}, \mathbf{R}, t)$ describes the situation where the center of mass of the two-particle pair is at $\mathbf{r}$, with separation distance vector **R**. Without loss of generality, this function can be reexpressed as

$$\tilde{n}_2(\mathbf{r}, \mathbf{R}, t) = n\left(\mathbf{r} + \frac{1}{2}\mathbf{R}, t\right) n\left(\mathbf{r} - \frac{1}{2}\mathbf{R}, t\right) \tilde{g}(\mathbf{r}, \mathbf{R}, t) \tag{5.44}$$

where $\tilde{g}(\mathbf{r}, \mathbf{R}, t)$ is called a nonequilibrium radial distribution function, and $n(\mathbf{r} + \frac{1}{2}\mathbf{R})$ and $n(\mathbf{r} - \frac{1}{2}\mathbf{R})$ are the single molecule density functions. Expanding the single molecule density functions in a Taylor series about $\mathbf{r}$ [cf. Eq. (5.43)] and neglecting second-order and higher derivatives, Eq. (5.44) becomes

$$\tilde{n}_2(\mathbf{r}, \mathbf{R}, t) \cong n^2(\mathbf{r}, t)\tilde{g}(\mathbf{r}, \mathbf{R}, t) \tag{5.45}$$

Finally, substituting Eqs. (5.43) and (5.45) into Eq. (5.37) gives

$$\int n_2(\mathbf{r}, \mathbf{r}', t)\frac{\partial \Phi_{\text{int}}(\mathbf{R})}{\partial \mathbf{R}} d\mathbf{R} = n^2(\mathbf{r}, t)\int \tilde{g}(\mathbf{r}, \mathbf{R}, t)\frac{\partial \Phi_{\text{int}}(\mathbf{R})}{\partial \mathbf{R}} d\mathbf{R}$$
$$+ \frac{\partial}{\partial \mathbf{r}} \cdot \int \mathbf{R}\frac{1}{2}n^2(\mathbf{r}, t)\tilde{g}(\mathbf{r}, \mathbf{R}, t)\frac{\partial \Phi_{\text{int}}(\mathbf{R})}{\partial \mathbf{R}} d\mathbf{R} \tag{5.46}$$

With the symmetry properties of the two-particle distribution

$$\tilde{g}(\mathbf{r}, \mathbf{R}, t) = \tilde{g}(\mathbf{r}, -\mathbf{R}, t) \tag{5.47}$$

and from Newton's third law

$$\frac{\partial \Phi(\mathbf{R})}{\partial \mathbf{R}} = -\frac{\partial \Phi(-\mathbf{R})}{\partial(-\mathbf{R})} \tag{5.48}$$

it can be seen that the integrand of the first integral in Eq. (5.46) is antisymmetric with respect to the sign of **R** and therefore the integral vanishes. The second integral can be written as

$$\frac{\partial}{\partial \mathbf{r}} \cdot \int \mathbf{R}\frac{1}{2}n^2(\mathbf{r}, t)\tilde{g}(\mathbf{r}, \mathbf{R}, t)\frac{\partial \Phi_{\text{int}}(\mathbf{R})}{\partial \mathbf{R}} d\mathbf{R}$$
$$= \frac{1}{2}\frac{\partial}{\partial \mathbf{r}} \cdot n^2(\mathbf{r}, t)\int \tilde{g}(\mathbf{r}, \mathbf{R}, t)\frac{\mathbf{R}\mathbf{R}}{R}\frac{d\,\Phi_{\text{int}}(R)}{dR} d\mathbf{R} \tag{5.49}$$

where, for the sake of simplicity, we have assumed a spherically symmetric interaction potential (cf. Chap. 4, Sec. 7)

$$\frac{\partial \Phi(\mathbf{R})}{\partial \mathbf{R}} = \frac{\mathbf{R}}{R}\frac{d\,\Phi(R)}{dR} \tag{5.50}$$

Summarizing the results from Eqs. (5.22), (5.26), (5.35), and (5.49) gives

$$\frac{\partial}{\partial t}(\rho \mathbf{v}_0) + \frac{\partial}{\partial \mathbf{r}} \cdot (\rho \mathbf{v}_0 \mathbf{v}_0) = -\frac{\partial}{\partial \mathbf{r}} \cdot \mathbf{P} + \rho\frac{\mathbf{F}_{\text{ext}}}{m} \tag{5.51}$$

or, expanding the left-hand side and subtracting off the equation of continuity, Eq. (5.19), leads to the general form of the momentum conservation equation or the so-called *Equation of Motion* as

$$\rho\frac{\partial}{\partial t}\mathbf{v}_0 + \rho\left(\mathbf{v}_0\cdot\frac{\partial}{\partial \mathbf{r}}\right)\mathbf{v}_0 = -\frac{\partial}{\partial \mathbf{r}}\cdot\mathbf{P} + \rho\frac{\mathbf{F}_{\text{ext}}}{m} \tag{5.52}$$

where $\mathbf{P}$ is the so-called pressure tensor defined by

$$\mathbf{P} \equiv \mathbf{P}_k + \mathbf{P}_\phi \tag{5.53}$$

with

$$\mathbf{P}_k = m\int\left(\frac{\mathbf{p}}{m} - \mathbf{v}_0\right)\left(\frac{\mathbf{p}}{m} - \mathbf{v}_0\right)f_1(\mathbf{r},\mathbf{p},t)\,d\mathbf{p} \tag{5.54}$$

and

$$\mathbf{P}_\phi = -\frac{1}{2}n^2(\mathbf{r},t)\int\tilde{g}(\mathbf{r},\mathbf{R},t)\frac{\mathbf{RR}}{R}\frac{d\,\Phi_{\text{int}}(R)}{d\,R}d\mathbf{R} \tag{5.55}$$

As discussed in Chap. 4, under equilibrium conditions the function f_1 becomes a gaussian or maxwellian distribution function for the molecular velocities, and g becomes an equilibrium radial distribution function, i.e.,

$$\tilde{g}(\mathbf{r},\mathbf{R},t) = g(R) \qquad \text{(equilibrium only)} \tag{5.56}$$

Under this condition, Eq. (5.55) can be written as

$$\mathbf{P}_\phi^{\text{eq}} = -\frac{1}{2}n^2\int g(R)\frac{d\,\Phi_{\text{int}}}{d\,R}\delta_R\delta_R R\,d\mathbf{R} \qquad \text{(equilibrium only)} \tag{5.57}$$

which is Eq. (4.46), obtained previously, where δ_R is a radial unit vector in spherical coordinates ($\delta_R \equiv \mathbf{R}/R$). Using the relationship (see Chap. 4, footnotes)

$$\int_{\text{all }\mathbf{R}} f(R)\delta_R\delta_R d\mathbf{R} = \frac{4\pi}{3}\mathbf{I}\int_0^\infty R^2 f(R)d\,R \tag{5.58}$$

Eq. (5.57) can be rewritten as

$$\mathbf{P}_\phi^{\text{eq}} = -\frac{2}{3}\pi n^2\mathbf{I}\int_0^\infty g(R)\frac{d\,\Phi_{\text{int}}}{d\,R}R^3 d\,R \qquad \text{(equilibrium only)} \tag{5.59}$$

which is recognized as the integral term in the equation of state given previously in Chap. 4 [cf. Eq. (4.74)]. The pressure tensor for *nonequilibrium* systems, Eq. (5.53), will be treated extensively in Chap. 6. There it will be shown that a nonequilibrium pressure tensor expression can be approximately obtained from a perturbation expansion about a *local* equilibrium state.

5.5 The Energy Balance Equation

In order to obtain the energy conservation equation, Irving and Kirkwood[1] showed that α should be given by

$$\alpha = \frac{1}{2m}\sum_{k=1}^{N} p_k^2\delta(\mathbf{r}_k - \mathbf{r}) + \sum_{k=1}^{N}\phi_{\text{ext}}(\mathbf{r},t)\delta(\mathbf{r}_k - \mathbf{r})$$

$$+\frac{1}{2}\sum_{j}\sum_{\substack{k \\ k\neq j}}\phi_{\text{int}}(|\mathbf{r}_j - \mathbf{r}_k|)\delta(\mathbf{r}_k - \mathbf{r}) \tag{5.60}$$

Thus, α represents a sum of the kinetic, external potential (ϕ_{ext}), and intermolecular potential (ϕ_{int}) energies at the locator vector $\mathbf{r}$ and at a time t. Substituting Eq. (5.60) into Eq. (5.1), and following the same type of manipulations given in Secs. 5.3 and 5.4 leads to (Prob. 5.2)

$$\langle\alpha\rangle = n\left[\bar{U}_k(\mathbf{r},t) + \bar{U}_\phi(\mathbf{r},t) + \frac{1}{2}mv_0^2 + \phi_{\text{ext}}(\mathbf{r},t)\right] \tag{5.61}$$

where

$$\bar{U}_k(\mathbf{r},t) = \frac{m}{2n}\int\left(\frac{\mathbf{p}}{m} - \mathbf{v}_0\right)^2 f_1(\mathbf{r},\mathbf{p},t)d\mathbf{p} \tag{5.62}$$

is the average kinetic energy per molecule relative to the average velocity, and

$$\bar{U}_\phi(\mathbf{r},t) = \frac{1}{2}\int\phi_{\text{int}}(|\mathbf{r} - \mathbf{r}_1|)n_2(\mathbf{r},\mathbf{r}_1,t)d\mathbf{r}_1 \tag{5.63}$$

is the average intermolecular interaction potential energy per molecule. Now, as given in Chap. 4, the average internal energy per molecule $\bar{U}(\mathbf{r},t)$ is defined as the sum of the latter two quantities, i.e.,

$$\bar{U}(\mathbf{r},t) \equiv \bar{U}_k(\mathbf{r},t) + \bar{U}_\phi(\mathbf{r},t) \tag{5.64}$$

Now, substituting Eq. (5.60) into the first term on the right-hand side of Eq. (5.9) and simplifying gives (Prob. 5.3)

$$\frac{1}{N!}\frac{1}{m}\int\int\sum_i f_N\mathbf{p}_i\cdot\frac{\partial\alpha}{\partial\mathbf{r}_i}d\mathbf{r}^N d\mathbf{p}^N = -\frac{\partial}{\partial\mathbf{r}}\cdot(\mathbf{v}_0\cdot\mathbf{P}_k)$$

$$-\frac{\partial}{\partial\mathbf{r}}\cdot\mathbf{q}_k - \frac{\partial}{\partial\mathbf{r}}\cdot(n\mathbf{v}_0\bar{U}_k) - \frac{\partial}{\partial\mathbf{r}}\cdot\left[\mathbf{v}_0\left(\frac{1}{2}nmv_0^2\right)\right] - \frac{\partial}{\partial\mathbf{r}}\cdot(\phi_{\text{ext}}n\mathbf{v}_0)$$

$$-\frac{1}{2m}\frac{\partial}{\partial\mathbf{r}}\cdot\int\mathbf{j}_1^{(2)}(\mathbf{r},\mathbf{r}_j,t)\phi_{\text{int}}(|\mathbf{r}_j - \mathbf{r}|)d\mathbf{r}_j \tag{5.65}$$

where

$$\mathbf{q}_k \equiv \frac{1}{2m^2}\int f_1(\mathbf{r},\mathbf{p},t)|\mathbf{p}-m\mathbf{v}_0|^2(\mathbf{p}-m\mathbf{v}_0)d\,\mathbf{p} \tag{5.66}$$

and

$$\mathbf{j}_1^{(2)}(\mathbf{r}_i,\mathbf{r}_j,t) \equiv \int\int f_2(\mathbf{r}_i,\mathbf{r}_j,\mathbf{p}_i,\mathbf{p}_j)\mathbf{p}_i d\,\mathbf{p}_i d\,\mathbf{p}_j \tag{5.67}$$

Finally, substituting Eq. (5.60) into the second term on the right-hand side of Eq. (5.9) and simplifying gives

$$-\frac{1}{N!}\int\int\sum_{i=1}^{N}\left(f_N\frac{\partial\Phi}{\partial\mathbf{r}_i}\cdot\frac{\partial\alpha}{\partial\mathbf{p}_i}\right)d\,\mathbf{r}^N d\,\mathbf{p}^N$$

$$=\frac{1}{m}\int\mathbf{j}_1^{(2)}(\mathbf{r},\mathbf{r}_j,t)\frac{\partial\phi(|\mathbf{r}-\mathbf{r}_j|)}{\partial\mathbf{r}_j}d\,\mathbf{r}_j \tag{5.68}$$

Summarizing our results, the energy conservation equation up to this point is

$$\frac{\partial}{\partial t}n\left(\bar{U}_k+\frac{1}{2}mv_0^2+\bar{U}_\phi+\phi_{\text{ext}}\right)=-\frac{\partial}{\partial\mathbf{r}}\cdot(\mathbf{v}_0\cdot\mathbf{P}_k)-\frac{\partial}{\partial\mathbf{r}}\cdot\mathbf{q}_k$$
$$-\frac{\partial}{\partial\mathbf{r}}\cdot n\mathbf{v}_0\left(\bar{U}_k+\frac{1}{2}mv_0^2+\phi_{\text{ext}}\right)$$
$$+\frac{1}{m}\int\mathbf{j}_1^{(2)}(\mathbf{r},\mathbf{r}_j,t)\frac{\partial\phi_{\text{int}}(|\mathbf{r}-\mathbf{r}_j|)}{\partial\mathbf{r}_j}d\,\mathbf{r}_j$$
$$-\frac{1}{2m}\frac{\partial}{\partial\mathbf{r}}\cdot\int\mathbf{j}_1^{(2)}(\mathbf{r},\mathbf{r}_j,t)\phi_{\text{int}}(|\mathbf{r}_j-\mathbf{r}|)d\,\mathbf{r}_j \tag{5.69}$$

As in the derivation of the momentum equation, we can transfer to center of mass and relative coordinates and expand $\mathbf{j}_1^{(2)}$ as [cf. Eq. (5.43)]

$$\mathbf{j}_1^{(2)}(\mathbf{r},\mathbf{r}_j,t)=\tilde{\mathbf{j}}_1^{(2)}(\mathbf{r}_{\text{cm}},\mathbf{R},t)=\tilde{\mathbf{j}}_1^{(2)}(\mathbf{r},\mathbf{R},t)+\frac{1}{2}\mathbf{R}\cdot\frac{\partial}{\partial\mathbf{r}}\tilde{\mathbf{j}}_1^{(2)}(\mathbf{r},\mathbf{R},t)+\cdots \tag{5.70}$$

Now, with the symmetry property [cf. Eq. (5.47)]

$$\tilde{\mathbf{j}}_1^{(2)}(\mathbf{r}_{\text{cm}},\mathbf{R},t)=\tilde{\mathbf{j}}_1^{(2)}(\mathbf{r}_{\text{cm}},-\mathbf{R},t) \tag{5.71}$$

the integral terms in Eq. (5.69) simplify, leading to

$$\frac{\partial}{\partial t}n\left(\bar{U}_k+\frac{1}{2}mv_0^2+\bar{U}_\phi+\phi_{\text{ext}}\right)=-\frac{\partial}{\partial\mathbf{r}}\cdot(\mathbf{v}_0\cdot\mathbf{P}_k)-\frac{\partial}{\partial\mathbf{r}}\cdot\mathbf{q}_k$$
$$-\frac{\partial}{\partial\mathbf{r}}\cdot n\mathbf{v}_0\left(\bar{U}_k+\frac{1}{2}mv_0^2+\phi_{\text{ext}}\right)$$

$$+\frac{1}{2m}\frac{\partial}{\partial\mathbf{r}}\cdot\int\tilde{\mathbf{j}}_1^{(2)}(\mathbf{r},\mathbf{R},t)\cdot\frac{\mathbf{RR}}{R}\frac{\partial\phi_{\text{int}}(R)}{\partial R}d\mathbf{R}$$

$$-\frac{1}{2m}\frac{\partial}{\partial\mathbf{r}}\cdot\int\tilde{\mathbf{j}}_1^{(2)}(\mathbf{r},\mathbf{R},t)\phi_{\text{int}}(R)d\mathbf{R} \tag{5.72}$$

where we have neglected second-order and higher spatial gradient terms.

Now, to get the energy balance equation in its usual form, we add and subtract on the right-hand side

$$\frac{\partial}{\partial\mathbf{r}}\cdot(n\mathbf{v}_0\bar{U}_\phi)$$

and

$$\frac{\partial}{\partial\mathbf{r}}\cdot(\mathbf{v}_0\cdot\mathbf{P}_\phi)$$

leading to

$$\frac{\partial}{\partial t}n\left(\bar{U}+\frac{1}{2}mv_0^2+\phi_{\text{ext}}\right)+\frac{\partial}{\partial\mathbf{r}}\cdot n\mathbf{v}_0\left(\bar{U}+\frac{1}{2}mv_0^2+\phi_{\text{ext}}\right)$$

$$=-\frac{\partial}{\partial\mathbf{r}}\cdot(\mathbf{v}_0\cdot\mathbf{P})-\frac{\partial}{\partial\mathbf{r}}\cdot\mathbf{q} \tag{5.73}$$

where

$$\mathbf{q}=\mathbf{q}_k+\mathbf{q}_\phi \tag{5.74}$$

is the total *heat flux vector*, and the potential part $\mathbf{q}_\phi$ is defined as

$$\mathbf{q}_\phi\equiv\frac{1}{2m}\int\left[\phi_{\text{int}}(R)\mathbf{I}-\frac{\mathbf{RR}}{R}\frac{\partial\phi_{\text{int}}(R)}{\partial R}\right]$$

$$\times\left[\tilde{\mathbf{j}}_1^{(2)}(\mathbf{r},\mathbf{R},t)-m\mathbf{v}_0n^2\tilde{g}(\mathbf{r},\mathbf{R},t)\right]d\mathbf{R} \tag{5.75}$$

Note that the energy terms in parentheses in Eq. (5.73) (internal plus macrokinetic plus potential) represent the total macroscopic energy of the fluid and, thus, Eq. (5.73) is the general differential form of the first law of thermodynamics for *open systems*.

Finally, we can eliminate the mass and momentum conservation equations from Eq. (5.73) to write the energy balance equation, assuming for the sake of simplicity that ϕ_{ext} is time independent, as (Prob. 5.4)

$$n\frac{\partial}{\partial t}\bar{U}+n\mathbf{v}_0\cdot\frac{\partial}{\partial\mathbf{r}}\bar{U}=-\left(\frac{\partial}{\partial\mathbf{r}}\cdot\mathbf{q}\right)-\left(\mathbf{P}:\frac{\partial\mathbf{v}_0}{\partial\mathbf{r}}\right) \tag{5.76}$$

This equation is well-known in energy transport phenomena, where the left-hand side accounts for temporal and spatial changes in the internal energy per molecule $\bar{U}$, and the first and second term on the right-hand side represent the heat flux vector and *viscous dissipation* terms, respectively. Looking at Eqs. (5.66) and (5.75) it can be seen that $\mathbf{q}$ represents the transfer of the molecular kinetic energy and intermolecular interaction energies (a type of potential energy), *all relative to the bulk velocity* $\mathbf{v}_0$. Problem 5.5 examines how the energy conservation equation can be recast in terms of temperature.

In Chap. 6, we will examine various approaches for the determination of $\mathbf{P}$ and $\mathbf{q}$ for specific types of fluids. Finally, we complete our derivations of the transport equations by developing the equation for the conservation of entropy.

5.6 Entropy Conservation

In previous chapters, we have introduced two seemingly different definitions of entropy, i.e., Boltzmann's entropy, Eq. (3.41), celebrated in nonequilibrium studies

$$S_B(\mathbf{r},t) = -\frac{k}{m}\int f_1(\mathbf{r},\mathbf{p},t)\ln\, f_1(\mathbf{r},\mathbf{p},t)d\,\mathbf{p}$$

and Gibbs entropy, Eq. (4.39), celebrated in equilibrium systems

$$S_G = -\frac{1}{N!h^{3N}}\int\!\!\int f'_N \ln\, f'_N d\,\mathbf{r}^N d\,\mathbf{p}^N$$

where $f'_N = h^{3N} f_N$. We have seen that Boltzmann's entropy definition leads to the *law of entropy increase*, whereas it will be shown in this chapter that the Gibbs entropy, as defined by Eq. (4.39), does not. Which definition, or their modifications, is actually correct is still the subject of some scientific debate. However, we will show here that it is possible to introduce a more general definition of entropy that encompasses both Boltzmann's and Gibbs' ideas as limiting cases. This exercise will help us to pin point what specific characteristics of entropy, at least as defined here, lead to an entropy conservation equation that encompasses the law of entropy increase, and it will allow us to practically extend entropy analysis to nonequilibrium gas and liquid systems in a manner consistent with the equilibrium expressions introduced in Chap. 4.

We begin by defining a general quantity α as

$$\alpha = -k\left[\frac{N!}{s!\,(N-s)!}\right]\frac{1}{s}\sum_{j=1}^{s}\delta\,(\mathbf{r}_j - \mathbf{r})\ln\,[\hbar^{3s} f_s(\mathbf{r}^s,\mathbf{p}^s,t)] \qquad (5.77)$$

where f_s is an sth order, reduced density function, introduced in Chap. 3 [see Eq. (3.7)]. Substituting into Eq. (5.1), the average of α,

$\langle\alpha\rangle$, is given by

$$\langle\alpha\rangle = n\bar{S}_{\mathrm{BG}} = -\frac{k}{s!}\frac{1}{s}\sum_{j=1}^{s}\int\int \delta\,(\mathbf{r}_j - \mathbf{r})\, f_s \ln(\hbar^{3s} f_s)\, d\,\mathbf{r}^s d\,\mathbf{p}^s \tag{5.78}$$

We have called this average the Boltzmann-Gibbs entropy, where $\bar{S}_{\mathrm{BG}}$ is the entropy per molecule and n is the molecular number density. For $s = 1$, we recover Boltzmann's definition

$$n\bar{S}_B(\mathbf{r}, t) = -k\int f_1(\mathbf{r}, \mathbf{p}', t)\ln(\hbar^3 f_1)\, d\,\mathbf{p}' \tag{5.79}$$

and for $s = N$ we obtain Gibbs entropy

$$n\bar{S}_G(\mathbf{r}, t) = -\frac{k}{N!}\frac{1}{N}\sum_{j=1}^{N}\int\int \delta\,(\mathbf{r}_j - \mathbf{r})\, f_N \ln(\hbar^{3N} f_N)\, d\,\mathbf{r}^N d\,\mathbf{p}^N \tag{5.80}$$

We note that for systems at equilibrium, f_N is independent of the locator vector $\mathbf{r}$ and Eq. (5.80) reduces to Gibbs entropy for equilibrium systems, Eq. (4.39).† Also, note that the introduction of Planck's constant in the logarithm term of Boltzmann's entropy, Eq. (5.79), is necessary on account of dimensional arguments, albeit it is often incorrectly left out.

Now, to obtain an entropy conservation equation, we can work with Eq. (5.9) modified to include the time dependence in α itself, or it is somewhat easier to work directly with the reduced Liouville equation, Eq. (3.20) or (3.24), for pairwise additive systems; we choose the latter representation.

Multiplying Eq. (3.24) by

$$-\frac{k}{s!}\frac{1}{s}\sum_{j=1}^{s}\ln(\hbar^{3s} f_s)\delta(\mathbf{r}_j - \mathbf{r})$$

and integrating over all $(\mathbf{r}^s, \mathbf{p}^s)$ space gives the following term by term analysis. For the time dependent term, we have

$$\int\int\left(-\frac{k}{s!s}\right)\sum_{j=1}^{s}\ln(\hbar^{3s} f_s)\delta(\mathbf{r}_j - \mathbf{r})\frac{\partial f_s}{\partial t}d\,\mathbf{r}^s d\,\mathbf{p}^s$$

$$= \frac{\partial}{\partial t}\int\int\left(-\frac{k}{s!s}\right)\sum_{j=1}^{s} f_s\ln(\hbar^{3s} f_s)\delta(\mathbf{r}_j - \mathbf{r})\, d\,\mathbf{r}^s d\,\mathbf{p}^s$$

†Also, note $n \equiv N/V = N/\int d\,\mathbf{r}$ at equilibrium.

$$-\frac{k}{s!s}\sum_{j=1}^{s}\int\int f_s\frac{\partial}{\partial t}[\ln(\hbar^{3s}f_s)]\delta(\mathbf{r}_j-\mathbf{r})\,d\mathbf{r}^s d\mathbf{p}^s$$

$$=\frac{\partial}{\partial t}(n\bar{S}_{\text{BG}})+\frac{k}{s}\frac{\partial n(\mathbf{r},t)}{\partial t} \tag{5.81}$$

For the second term in Eq. (3.24), we can write

$$-\frac{k}{s!s}\sum_{i=1}^{s}\sum_{j=1}^{s}\int\int\left(\frac{\mathbf{p}_i}{m_i}\cdot\frac{\partial f_s}{\partial \mathbf{r}_i}\right)\ln(\hbar^{3s}f_s)\delta(\mathbf{r}_j-\mathbf{r})\,d\mathbf{r}^s d\mathbf{p}^s$$

$$=-\frac{k}{s!s}\sum_{i=1}^{s}\sum_{j=1}^{s}\left\{\int\int\frac{\mathbf{p}_i}{m_i}\cdot\frac{\partial}{\partial \mathbf{r}_i}[f_s\ln(\hbar^{3s}f_s)]\delta(\mathbf{r}_j-\mathbf{r})\,d\mathbf{r}^s d\mathbf{p}^s\right.$$

$$\left.-\int\int\frac{\mathbf{p}_i}{m_i}\cdot f_s\frac{\partial}{\partial \mathbf{r}_i}[\ln(\hbar^{3s}f_s)]\delta(\mathbf{r}_j-\mathbf{r})\,d\mathbf{r}^s d\mathbf{p}^s\right\} \tag{5.82}$$

For $j \neq i$, the first term on the right-hand side of Eq. (5.82) vanishes by the virtue of Gauss' theorem and the properties[3]

$$\left.\begin{matrix} f_s \\ f_s \ln f_s \end{matrix}\right\} \to 0 \qquad \text{as } \mathbf{r}_i \to \pm\infty \tag{5.83}$$

Thus, the right-hand side of Eq. (5.82) reduces to

$$-\frac{k}{s!s}\frac{\partial}{\partial \mathbf{r}}\cdot\sum_{i=1}^{s}\int\int\frac{\mathbf{p}_i}{m_i}f_s\ln(\hbar^{3s}f_s)\,d\mathbf{r}^{s-1}d\mathbf{p}^s$$

$$+\frac{k}{s!}\sum_{i=1}^{s}\sum_{j=1}^{s}\int\int\frac{\mathbf{p}_i}{m_i}\cdot\frac{\partial f_s}{\partial \mathbf{r}_i}\delta(\mathbf{r}_j-\mathbf{r})\,d\mathbf{r}^s d\mathbf{p}^s \tag{5.84}$$

Substituting $\mathbf{p}_i/m_i = \mathbf{p}'_i/m_i + \mathbf{v}_0$ in the first term on the right-hand side of (5.84), where $\mathbf{p}'_i = \mathbf{p}_i - m\mathbf{v}_0$ is the molecular momentum relative to $\mathbf{v}_0$, and reducing the second term using the definition of $\mathbf{v}_0$ from Eq. (5.15) gives

$$\frac{\partial}{\partial \mathbf{r}}\cdot n\bar{S}_{\text{BG}}\mathbf{v}_0+\frac{\partial}{\partial \mathbf{r}}\cdot\mathbf{s}+\frac{k}{s}\left(\frac{\partial}{\partial \mathbf{r}}\cdot n\mathbf{v}_0\right) \tag{5.85}$$

where we have introduced the entropy flux vector $\mathbf{s}$ defined by

$$\mathbf{s}(\mathbf{r},t)\equiv-\frac{k}{s!s}\sum_{i=1}^{s}\int\int\frac{\mathbf{p}'_i}{m_i}f_s\ln(\hbar^{3s}f_s)\,d\mathbf{r}^{s-1}d\mathbf{p}^s \tag{5.86}$$

which is the flux of entropy relative to the mass average velocity $\mathbf{v}_0$. Note that the last term in Eq. (5.85) cancels with the last term in Eq. (5.81), by virtue of the continuity equation, Eq. (5.18).

Now it can be readily shown from Gauss' theorem that the third term in Eq. (3.24) integrates to zero (Prob. 5.4). Finally, the last term in Eq. (3.24) becomes

$$-\frac{k}{s!s}\sum_{i=1}^{s}\sum_{j=1}^{s}\int\int \ln(\hbar^{3s} f_s)\delta(\mathbf{r}_j - \mathbf{r})$$
$$\times\left[\frac{\partial\Phi(\mathbf{r}_{i,s+1})}{\partial\mathbf{r}_i}\cdot\frac{\partial f_{s+1}}{\partial\mathbf{p}_i}\right] d\mathbf{r}^s d\mathbf{r}_{s+1} d\mathbf{p}^s d\mathbf{p}_{s+1} \equiv n\bar{s}_g \qquad (5.87)$$

where we call $\bar{s}_g$ the entropy generation per molecule.

For $s = 1$, Eq. (5.87) becomes Boltzmann's entropy generation term, discussed in Chap. 3,

$$n\bar{s}_g(\mathbf{r},t) = -k\int\int \ln[\hbar^3 f_1(\mathbf{r},\mathbf{p}_1,t)]$$
$$\times\left[\frac{\partial\Phi(|\mathbf{r}-\mathbf{r}_2|)}{\partial\mathbf{r}_2}\cdot\frac{\partial f_2(\mathbf{r},\mathbf{r}_2,\mathbf{p}_1,\mathbf{p}_2,t)}{\partial\mathbf{p}_1}\right] d\mathbf{r}_2\, d\mathbf{p}_1\mathbf{p}_2 \qquad (5.88)$$

Equation (5.87) is a generalization of Boltzmann's entropy generation, Eq. (5.88), where entropy is defined in an s-space and the system dynamics are described in a higher $(s+1)$-space. Note that if the entropy had been defined in the $(s+1)$-space or the system dynamics described in the s-space, then the integral term of Eq. (5.87) would identically vanish by Gauss' theorem and the properties given by Eq. (5.83). The Gibbs entropy, defined in the total N-space, represents such a case, i.e., the system dynamics and entropy are both defined in the N-space. Thus, it appears that a *necessary* condition for entropy generation to exist, according to the definition here, is whenever the number of variables or degrees of freedom s as specified by $f_s \ln f_s$ is less than the number of variables required to completely specify the dynamics of the system. In the subject of information theory and statistics,[5] the entropy definition, Eq. (5.78) is also known as a *measure of uncertainty*. When this measure of uncertainty *misses* important dynamic variables (*hidden* variables), entropy as defined in the s-space may be generated.[†] In Boltzmann's case, the system dynamics are described in the $(s = 2)$-space and the entropy is defined in the lower, $(s = 1)$-space, which potentially *misses* important information about the system dynamics leading to the possibility of entropy or *uncertainty* generation.

[†]A corollary to this theorem is that any system with an observed increase in entropy accurately determined to include all variables known about the system and which is not believed to be interacting with any other system or its surroundings must contain hidden or missing variables that are not well described.

Although not proven here, Eq. (5.87) or (5.88) may be zero in a more general sense if all variables and distribution functions involved are accurately determined without any approximations or truncations. Furthermore, we still have *not* shown that the general form of the entropy generation term is *sufficient* to guarantee nonzero, positive entropy generation. In the case of Boltzmann's analysis, further approximations of Eq. (5.88) were shown in Chap. 3 to lead to the law of entropy increase for dilute gases. More generally, truncations of the BBKGY hierarchy, discussed in Chap. 3, are practically necessary for resolving higher-order distribution functions. E. T. Jaynes[4] has argued that only through such deliberate truncations and/or approximations is it possible to obtain positive entropy generation, which establishes a *sufficient* condition for positive, nonzero entropy generation.

Regardless of the arguments as to the exact nature of positive, nonzero entropy generation, the above analysis provides a general, practical framework on which entropy generation may be developed for both gas and liquid systems. More importantly, it allows for a bridge between equilibrium entropy analysis, such as Green's dense gas and liquid state entropy defined in the $(s = 2)$-space introduced in Chap. 4, and nonequilibrium entropy introduced in this chapter. Specifically, the nonequilibrium, two-space Green's entropy would be defined according to Eq. (5.78) as

$$n\bar{S}_{\text{Green}}(\mathbf{r}, t) = -\frac{k}{4}\sum_{j=1}^{2}\int\int \delta(\mathbf{r}_j - \mathbf{r}) f_2 \ln(\hbar^6 f_2)\, d\mathbf{r}^2 d\mathbf{p}^2 \tag{5.89}$$

which under equilibrium conditions, [Eq. (4.34)], can readily be shown to reduce to Green's expression given in Chap. 4, Eq. (4.68).† We will return to the specific approximations and truncations that lead to the law of entropy increase in Chap. 6 when "closure" of the transport equations are discussed.

Summarizing the results of Eqs. (5.81), (5.85), and (5.87), we have the following general entropy conservation equation

$$\frac{\partial(n\bar{S}_{\text{BG}})}{\partial t} + \frac{\partial}{\partial \mathbf{r}} \cdot (n\bar{S}_{\text{BG}}\mathbf{v}_0) = -\frac{\partial}{\partial \mathbf{r}} \cdot \mathbf{s} + n\bar{s}_g \tag{5.90}$$

†See Prob. 5.6 at the end of this chapter. Thus, the generalized Boltzmann-Gibbs entropy expression, Eq. (5.78), along with higher-order equilibrium expressions, provides a very simple and straightforward method of obtaining higher-order expressions for equilibrium entropy, introduced in Chap. 4, in addition to their fully consistent nonequilibrium counterparts given in this chapter.

Expanding the left-hand side and eliminating the equation of continuity, we get the equivalent form of the entropy conservation as

$$n\frac{\partial \bar{S}_{\mathrm{BG}}}{\partial t} + n\mathbf{v}_0 \cdot \frac{\partial \bar{S}_{\mathrm{BG}}}{\partial \mathbf{r}} = -\frac{\partial}{\partial \mathbf{r}} \cdot \mathbf{s} + n\bar{s}_g \tag{5.91}$$

where $\mathbf{s}$ and $\bar{s}_g$ are defined by Eqs. (5.86) and (5.87), respectively.

5.7 Local Equilibrium Flows

In order to solve the conservation or transport equations (mass, momentum, energy, and entropy) in terms of the dependent variables n, $\mathbf{v}_0$, $\hat{U}$, and $\hat{S}$, we must further resolve the expressions for the flux vectors—$\mathbf{P}$, $\mathbf{q}$, and $\mathbf{s}$ and entropy generation $\bar{s}_g$. This resolution is the subject of "closure," which will be treated in some detail in the next chapter. However, as a matter of illustration and for future reference, we can resolve the flux vector expression for what is called the *local equilibrium approximation*, i.e., we assume that the N-molecule distribution function locally follows the equilibrium form developed in Chap. 4, i.e., we write [cf. Eq. (4.34)]

$$f_N{}^{\text{loc eq}} = \frac{n_N(\mathbf{r}^N, t)}{[2\pi mkT\,(\mathbf{r}, t)]^{3N/2}} \exp\left\{ -\sum_{j=1}^{N} \frac{[\mathbf{p}_j - \mathbf{v}_0(\mathbf{r}, t)]^2}{2mkT\,(\mathbf{r}, t)} \right\} \tag{5.92}$$

where $n_N(\mathbf{r}^N, t)$ is the N-particle equilibrium configurational distribution function, which is now possibly time dependent through the time dependence of the local number density $n(\mathbf{r}, t)$, local temperature $T = T\,(\mathbf{r}, t)$, and/or local velocity $\mathbf{v}_0 = \mathbf{v}_0(\mathbf{r}, t)$, where $\mathbf{r}$ is the locator vector in the laboratory frame.

Now, let's look at each flux term utilizing the distribution function, Eq. (5.92). We require both the first-order and second-order distribution functions, which under local equilibrium conditions follow Eq. (5.92) as [see Eqs. (4.47), (4.66), and (4.72)]

$$f_2(\mathbf{r}_1, \mathbf{r}_2, \mathbf{p}_1, \mathbf{p}_2, t) = \frac{n_2(\mathbf{r}_1, \mathbf{r}_2, t)}{[2\pi mkT\,(\mathbf{r}_1, t)]^3} \exp\left[-\sum_{j=1}^{2} \frac{(\mathbf{p}_j - \mathbf{v}_0)^2}{2mkT\,(\mathbf{r}_1, t)} \right] \tag{5.93}$$

$$n_2(\mathbf{r}_1, \mathbf{r}_2, t) = n^2(\mathbf{r}_1, t)\, g(r_{12}) \tag{5.94}$$

$$f_1(\mathbf{r}_1, \mathbf{p}_1, t) = \frac{n(\mathbf{r}_1, t)}{[2\pi mkT\,(\mathbf{r}_1, t)]^{\frac{3}{2}}} \exp\left[-\frac{(\mathbf{p}_1 - \mathbf{v}_0)^2}{2mkT\,(\mathbf{r}_1, t)} \right] \tag{5.95}$$

where $g(r_{12})$ is the equilibrium pair distribution function given, in pairwise additive systems, by the integral equation, Eq. (4.115).

Thus, the pressure tensor is directly obtained from Eqs. (5.53) and (5.54) as [cf. Eqs. (4.73) and (4.74)]

$$\mathbf{P} = p\mathbf{I} \qquad \text{(local equilibrium approximation)} \tag{5.96}$$

where

$$p = n(\mathbf{r}, t)kT(\mathbf{r}, t) - \frac{1}{6}[n(\mathbf{r}, t)]^2 \int_0^\infty g(R; T)4\pi R^3 \frac{d\Phi(R)}{dR} dR \tag{5.97}$$

and $R \equiv r_{12}$. Note in the last equation we have explicitly written $\Phi(R; T)$ and $g(R; T)$ due to the possible temperature dependence of the intermolecular interaction potential. Also, as discussed in Chap. 4, we have assumed a spherically symmetric interaction potential function.

Now, by inspection it can be seen that under local equilibrium conditions the integrals defining $\mathbf{q}^k, \mathbf{j}^{(2)}, \mathbf{s}$, and $\bar{s}_g$ [Eqs. (5.66), (5.67), (5.86), and (5.87), respectively] are antisymmetric with respect to the sign of $\mathbf{p}$ and therefore must vanish. Thus

$$\left.\begin{aligned} \mathbf{q} &= 0 \\ \mathbf{s} &= 0 \\ \bar{s}_g &= 0 \end{aligned}\right\} \text{local equilibrium approximation} \tag{5.98}$$

Finally, we can summarize the transport or conservation equations [Eqs. (5.19), (5.52), (5.73), and (5.90)] under the local equilibrium approximation as

$$\text{mass} \qquad \frac{\partial \rho}{\partial t} + \frac{\partial}{\partial \mathbf{r}} \cdot (\rho \mathbf{v}_0) = 0 \tag{5.99}$$

$$\text{momentum} \qquad \frac{\partial \mathbf{v}_0}{\partial t} + \left(\mathbf{v}_0 \cdot \frac{\partial}{\partial \mathbf{r}}\right)\mathbf{v}_0 = -\frac{1}{\rho}\frac{\partial p}{\partial \mathbf{r}} + \frac{\mathbf{F}_{\text{ext}}}{m} \tag{5.100}$$

$$\text{energy} \qquad n\frac{\partial \bar{U}}{\partial t} + n\mathbf{v}_0 \cdot \frac{\partial \bar{U}}{\partial \mathbf{r}} = -p\left(\frac{\partial}{\partial \mathbf{r}} \cdot \mathbf{v}_0\right) \tag{5.101}$$

$$\text{entropy} \qquad n\frac{\partial \bar{S}_{\text{BG}}}{\partial t} + n\mathbf{v}_0 \cdot \frac{\partial \bar{S}_{\text{BG}}}{\partial \mathbf{r}} = 0 \tag{5.102}$$

where p is given by a thermodynamic equation of state, Eq. (5.97), $p = p(n, T)$, which depends on the specific form of the intermolecular interaction potential.

Example 5.1 Isothermal Compressible Flow For local equilibrium, isothermal flows in the absence of external forces, the conservation of mass and momentum are, respectively,

$$\frac{\partial \rho}{\partial t} + \frac{\partial}{\partial \mathbf{r}} \cdot (\rho \mathbf{v_0}) = 0 \tag{5.103}$$

$$\frac{\partial \mathbf{v_0}}{\partial t} + \left(\mathbf{v_0} \cdot \frac{\partial}{\partial \mathbf{r}}\right) \mathbf{v_0} = -\frac{1}{\rho}\left(\frac{\partial \rho}{\partial \mathbf{r}}\right)\left(\frac{\partial p}{\partial \rho}\right)_T \tag{5.104}$$

where in this last equation we have used the chain rule under the isothermal condition

$$\frac{\partial p}{\partial \mathbf{r}} = \frac{\partial \rho}{\partial \mathbf{r}}\left(\frac{\partial p}{\partial \rho}\right)_T \tag{5.105}$$

Now, consider a one-dimensional system so that Eqs. (5.103) and (5.104) can be written as

$$\frac{\partial \rho}{\partial t} + \frac{\partial}{\partial z}(\rho v_{z_0}) = 0 \tag{5.106}$$

$$\rho\left[\frac{\partial v_{z_0}}{\partial t} + v_{z_0}\frac{\partial v_{z_0}}{\partial z}\right] = -c^2\frac{\partial \rho}{\partial z} \tag{5.107}$$

where

$$c \equiv \sqrt{(\partial p/\partial \rho)_T} \tag{5.108}$$

which is known from the equation of state of the fluid. In principle, Eqs. (5.95) and (5.96) can be solved simultaneously to yield $\rho(z,t)$ and $\mathrm{v}_{z_0}(z,t)$. Since Eq. (5.96) is nonlinear, such a solution is not straightforward, and here we seek to examine a linearized version of the problem. Consider the density to be composed of a uniform value ρ_0 plus a smaller disturbance quantity $\rho'(z,t)$. We seek to determine ρ' and the associated disturbance velocity $\mathrm{v}'(z,t)$, thus we write

$$\rho = \rho_0 + \rho' \tag{5.109}$$

$$v_{z_0} = v' \tag{5.110}$$

Substituting into Eqs. (5.106) and (5.107) and retaining terms through first-order in disturbance quantities gives

$$\text{mass} \qquad \frac{\partial \rho'}{\partial t} + \rho_0\frac{\partial v'}{\partial z} = 0 \tag{5.111}$$

$$\text{momentum} \qquad \rho_0\frac{\partial v'}{\partial t} = -c^2\frac{\partial \rho'}{\partial z} \tag{5.112}$$

Now, taking the $\partial/\partial t$ of Eq. (5.99) and combining with the $\partial/\partial z$ of Eq. (5.100) leads to the single equation for ρ'

$$\frac{\partial^2 \rho'}{\partial t^2} - c^2 \frac{\partial^2 \rho'}{\partial z^2} = 0 \tag{5.113}$$

which is recognized as the so-called *wave equation*. The solution to Eq. (5.113) can readily demonstrated to be[6]

$$\rho' = \begin{cases} F\left(t - \frac{z}{c}\right) & t \geq \frac{z}{c} \\ 0 & t < \frac{z}{c} \end{cases} \tag{5.114}$$

where $F(t) = \rho'(0, t)$ is the initial disturbance function and c represents the propagation velocity of the disturbance (isothermal speed of sound). In terms of the virial equation of state developed in Chap. 4, we have from Eq. (4.75) (with $M/\hat{V} = \rho$; $M \equiv$ molecular weight)

$$p = \frac{RT}{M}\left[\rho + \frac{B(T)}{M}\rho^2\right] \tag{5.115}$$

and, therefore

$$c = \left[\frac{RT}{M} + \frac{2B(T)}{M}\rho\right]^{1/2} \tag{5.116}$$

(see Prob. 5.7)

Interlude 5.2 Applications and Practical Importance of Chemical Transport Phenomena The goals of the analysis of the chemical transport phenomena are to determine the spatial and temporal changes in the chemical thermodynamic properties, $\hat{V}$, $\mathbf{P}$, $\hat{U}$ or T, and $\hat{S}$ (density, pressure, energy or temperature, and entropy, respectively) for any particular system. Predicting the numerical values of these variables enables us to determine the system's practical performance and efficiency. For example, solutions to the mass and momentum conservation equations lead to numerical values of $\hat{V}$, $\mathbf{P}$, and $\mathbf{v}_0$, which give us information on the volumetric throughput rates and pressure tensor changes within the system. The pressure tensor field can be used to determine the net force and torque acting on solid surfaces either contained within or bounding the system through the fundamental relationships[7,8]

$$\mathbf{F}_{\text{surface}} = -\int_S \mathbf{P} \cdot \hat{n}\, dS$$

$$\mathbf{T}_{0\ \text{surface}} = -\int_S (\mathbf{r}_0 \times \mathbf{P}) \cdot \hat{n}\, dS$$

where $\hat{n}$ is a unit normal pointing outward from the surface area element dS (see Fig. 5.1). The first relationship above physically represents the net fluid momentum flux as measured by $\mathbf{P}$ across any particular surface S. Since, dimensionally, the momentum flux [mass × velocity/(area × time] is equivalent to a force per unit area by Newton's law, this momentum transfer can

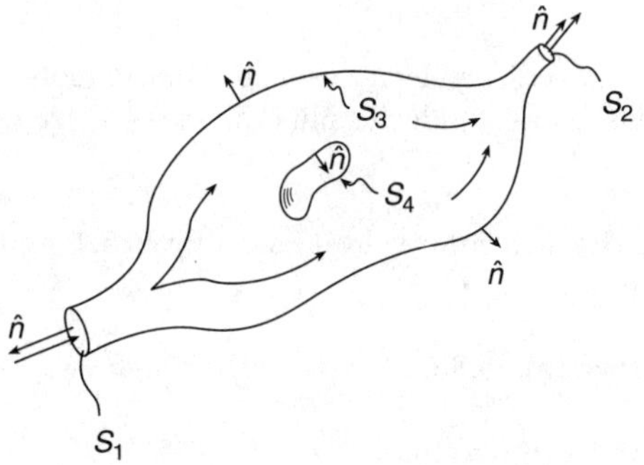

Figure 5.1 Illustration of the integrations of the property flux terms over a surface.

be interpreted as a force. This force will cause a movement of the surface unless the surface is acted upon by an equal and opposite restraining force. Similarly, the second expression involving the torque $\mathbf{T}_{0\,\text{surface}}$ represents the net fluid angular momentum flux as measured by $\mathbf{r}_0 \times \mathbf{P}$ where $\mathbf{r}_0$ locates a particular point on the surface relative to a fixed point "0." This torque will cause a rotation of the surface, relative to the point "0," unless the surface is restrained from doing so by an external restraining torque.

For the energy and entropy conservation equations, we can write similar and equally important expressions, viz.,

$$Q = -\int_S \mathbf{q} \cdot \hat{n}\, dS$$

$$W = -\int_S (\mathbf{v} \cdot \mathbf{P}) \cdot \hat{n}\, dS$$

$$E = -\int_S \mathbf{s} \cdot \hat{n}\, dS$$

where the first equation defines the so-called heat transfer Q, which physically represents the net energy transfer as measured by $\mathbf{q}$ across any particular surface S. The second relationship defines the thermodynamic work done by the system W on its surroundings. Note that the negative signs are by the standard convention—energy or work added to the system is positive, and energy or work released to the surroundings is negative. The third expression defines the net entropy transfer E across the surface S as measured by $\mathbf{s}$. Solutions to momentum and energy conservation equations lead to expressions for the flux quantities $\mathbf{P}$, $\mathbf{q}$, and $\mathbf{s}$, which are subsequently utilized to determine W, Q, and E; these latter quantities are of fundamental importance in engineering analysis of biological, chemical, mechanical, and electrical systems.[†] It is important to point out that the well-known thermodynamic quantities heat (Q) and work (W) are not abstract quantities; rather they are rigorously defined by the relationships shown here.

[†]For specific examples and problems see, Ref. 9.

Problems

5.1 Prove Eq. (5.26) by writing the molecular velocity as a sum of the molecular velocity relative to the mass average velocity plus the mass average velocity.

5.2 Using the same substitution for the molecular velocity as in Prob. 5.1, and carrying out integrations, prove Eq. (5.6).

5.3 With α defined by Eq. (5.60), prove Eq. (5.65).

5.4 By multiplying the third term in Eq. (3.24) by

$$-\frac{k}{s!s}\sum_{j=1}^{s}\ln(\hbar^{3s}f_s)\delta(\mathbf{r}_j-\mathbf{r}),$$

integrating over all $(\mathbf{r}^s, \mathbf{p}^s)$ space, and using Gauss' theorem, show that the integral vanishes.

5.5 The energy conservation equation in terms of the total derivative d/dt can be written as

$$n\frac{d\bar{U}}{dt} = -(\nabla\cdot\mathbf{q}) - (\mathbf{P}:\nabla\mathbf{v})$$

where $\nabla \equiv \partial/\partial\mathbf{r}$ and[7]

$$\frac{d}{dt} \equiv \frac{\partial}{\partial t} + \mathbf{v}\cdot\frac{\partial}{\partial\mathbf{r}}$$

which represents the total change in time seen by an observer moving with velocity $\mathbf{v}$. Now, using the chain rule

$$\frac{d\bar{U}}{dt} = \left(\frac{\partial\bar{U}}{\partial\bar{V}}\right)_T\frac{d\bar{V}}{dt} + \left(\frac{\partial\bar{U}}{\partial T}\right)_{\bar{V}}\frac{dT}{dt}$$

and the well-known thermodynamic relationship

$$\left(\frac{\partial\bar{U}}{\partial\bar{V}}\right)_T = \left[-p + T\left(\frac{\partial p}{\partial T}\right)_{\bar{V}}\right]$$

show that the energy conservation equation in terms of temperature can be written as

$$n\bar{C}_V\left[\frac{\partial T}{\partial t} + \mathbf{v}\cdot\nabla T\right] = -(\nabla\cdot\mathbf{q}) - (\mathbf{P}:\nabla\mathbf{v}) - \frac{1}{n}(\nabla\cdot\mathbf{v})\left[-p + T\left(\frac{\partial p}{\partial T}\right)_{\bar{V}}\right]$$

where $\bar{C}_V \equiv (\partial\bar{U}/\partial T)_{\bar{V}}$ is the definition of the heat capacity at constant specific volume.

5.6 Use Eq. (4.34) to show that Green's two-space nonequilibrium entropy, Eq. (5.89), can be readily reduced to its equilibrium counterpart, Eq. (4.68).

5.7 Calculate the isothermal speed of sound c for the Lennard-Jones fluids: N_2, CO_2, and C_2H_4 listed in Table 1.1. Consider two different densities both at 20°C: $\rho = 0.01$ g/cm^3 and $\rho = 0.10$ g/cm^3. How do your results compare with the ideal gas values, i.e., retaining only the first term in Eq. (5.117)?

References

1. J. H. Irving and J. G. Kirkwood, Statistical mechanical theory of transport processes IV. The equations of hydrodynamics, *J. Chem. Phys.*, **18**, 817, 1950.
2. L. Schwartz, *Theorie des Distributions, Actualites Scientigues et Industrielles*, Nos. 1091 and 1122, Hermans, Paris, 1950–1951, or D. S. Jones, *The Theory of Generalized Functions*, Cambridge University Press, Cambridge, 1982.
3. M. Abramowitz and I. A. Stegun, *Handbook of Mathematical Functions*, Dover, New York, 1972. [See Sec. 4.1.31 in particular].
4. E. T. Jaynes, Information theory and statistical mechanics, *Phys. Rev.*, **106**, 620; **108**, 171, 1957.
5. A. Katz, *Principles of Statistical Mechanics: The Information Theory Approach*, W. H. Freeman, San Francisco, 1967.
6. G. Arfken, *Mathematical Methods for Physicists*, 3rd ed., Academic Press, New York, 1985.
7. R. B. Bird, W. E. Stewart and E. N. Lightfoot, *Transport Phenomena*, Wiley, New York, 1960.
8. H. Brenner, *Low Reynolds Number Hydrodynamics*, Martinus Niehof, Boston, 1970.
9. S. I. Sandler, *Chemical and Engineering Thermodynamics*, 3rd ed., Wiley, New York, 1999.

Further Reading

John Gamble Kirkwood Collected Works, I. Oppenheim, (ed.), Gordon & Breach, New York, 1967. [Except for entropy conservation, most of the material of the chapter is based on Ref. 1. Kirkwood contributed heavily to numerous aspects of the statistical mechanics of equilibrium and nonequilibrium systems, including pioneering work in polymer kinetic theory, time correlations, and brownian motion. Some of the most significant papers of Kirkwood and colleagues are reproduced in this book.]

R. B. Bird, O. Hassager, R. C. Armstrong, and C. F. Curtiss, *Dynamics of Polymeric Liquids, Vol. 2 Kinetic Theory*, Wiley, New York, 1977. [The method of expansion of the interaction force terms using center of mass and relative coordinates used here is adapted from this work. In particular, see Chap. 14.]

Chapter

6

Closure of the Transport Equations

My name is Sylvester McMonkey McBean.
And I've heard of your troubles. I've heard you're unhappy.
But I can fix that. I'm the Fix-it-up Chappie.
I've come here to help you. I have what you need.
And my prices are low. And I work at great speed."

—THEODOR SEUSS GEISEL,
The Sneetches

6.1 Introduction: Complexities of Scales in Time and Space

The general equations of change given in the previous chapter show that the property flux vectors **P**, **q**, and **s** depend on the nonequilibrium behavior of the lower-order distribution functions g(**r**, **R**, t), $f_2(\mathbf{r}, \mathbf{r}', \mathbf{p}, \mathbf{p}', t)$, and $f_1(\mathbf{r}, \mathbf{p}, t)$. These functions are, in turn, obtained from solutions to the reduced Liouville equation (RLE) given in Chap. 3. Unfortunately, this equation is difficult to solve without a significant number of approximations. On the other hand, these approximate solutions have led to the theoretical basis of the so-called phenomenological laws, such as Newton's law of viscosity, Fourier's law of heat conduction, and Boltzmann's entropy generation, and have consequently provided a firm molecular, theoretical basis for such well-known equations as the Navier-Stokes' equation in fluid mechanics, Laplace's equation in heat transfer, and the second law of thermodynamics, respectively. Furthermore, theoretical expressions to quantitatively predict fluid transport properties, such as the coefficient of viscosity and thermal conductivity, in terms of the intermolecular potential functions for specific systems, are also a direct result of these approximate solution methods. Other types of flows where, for example, the Navier-Stokes equations do not hold, such as flows in rarefied gases and in shock waves, have also been treated through particular approximate solutions to the RLE.

Most approximate solutions to the RLE are based on asymptotic analyses, which exploit combinations of physical parameters that are small in magnitude in some relative scale sense, and this will introduce us to the field of nondimensionalization and scaling analysis. In this chapter, it will be demonstrated that what ultimately determines, for example, whether or not particular systems can be described in equilibrium or nonequilibrium states, and what specific mathematical analysis is applicable, are a number of distinct or characteristic scales of length and time. As a somewhat extreme example, a characteristic time scale to study the movement of the center of mass of a large glacier across a continent is quite large, on the order of hundreds to thousands of years or more (referred to as a *geological time scale*). On the other hand, the time scale to study chemical concentration changes or flux behavior along the narrow, solid-liquid interfacial boundary of a melting glacier is relatively small, on the order of seconds.

Understanding how to characterize and utilize length- and time-scale analysis for any particular system is a major goal of this chapter. We will see immediate differences and their consequences as we compare some of the most basic systems, specifically gases versus liquids. We will show how scaling differences in these two systems lead to entirely different mathematical analysis as we attempt to approximately resolve expressions for the lower-ordered distribution functions and resulting flux expressions (closure of the transport equations). So, we begin below by examining the so-called scaled, nondimensional form of the RLE.

Before proceeding, we note that the methodologies and "tools" developed in this chapter can be quite powerful and have been proven to be extremely useful in the study of all types of problems in science and engineering. Successful scientists and engineers have mastered scaling analysis as a means to reduce highly complex problems to more manageable forms. (See Interlude 6.1.)

6.2 Scaling of the Reduced Liouville Equation

In order to nondimensionalize the reduced Liouville equation, we introduce the following dimensionless variables denoted by an asterisk:

$$t^* = t/t_0 \tag{6.1}$$

$$\mathbf{r}_i^* = \mathbf{r}_i/l_0 \tag{6.2}$$

$$\mathbf{p}_i^* = \mathbf{p}_i/p_0 \tag{6.3}$$

$$f_s^* = f_s p_0^{3s} n_0^{-s} \tag{6.4}$$

$$f_{s+1}^* = f_{s+1} p_0^{3(s+1)} n_0^{-(s+1)} \tag{6.5}$$

$$\mathbf{F}_i^* = \mathbf{F}_i/F_0 = \mathbf{F}_i \lambda_0/\phi_0 \tag{6.6}$$

where $t_0, l_0, p_0(\equiv mv_0)$, and $F_0(\equiv \phi_0/\lambda_0)$ are characteristic values of time, length, momentum, and intermolecular interaction force, respectively, where the characteristic force is expressed as a characteristic interaction energy ϕ_0 divided by a characteristic interaction length scale λ_0. Substituting these dimensionless quantities into Eq. (3.24) via the chain rule, e.g.,

$$\frac{\partial f_s}{\partial t} = \frac{\partial f_s^*}{\partial t^*} \cdot \frac{dt^*}{dt} \cdot \frac{df_s}{df_s^*} = \left(\frac{v_0}{l_0}\right) \frac{n_0^s}{p_0^{3s}} \frac{\partial f_s^*}{\partial t^*} \tag{6.7}$$

and similarly for the spatial and momentum derivative terms, leads to the nondimensionless, reduced Liouville equation (Prob. 6.1)

$$\frac{\partial f_s^*}{\partial t^*} + \sum_{i=1}^{s} N_t \left[\mathbf{p}_i^* \cdot \frac{\partial f_s}{\partial \mathbf{r}_i^*} + \frac{1}{N_\alpha} \sum_{\substack{j=1 \\ j \neq i}}^{s} \mathbf{F}_i^* \left(\mathbf{r}_{ij}^*\right) \cdot \frac{\partial f_s^*}{\partial \mathbf{p}_i^*} \right]$$
$$= -N_t \frac{N_\beta}{N_\alpha} \sum_{i=1}^{s} \int\!\!\int \mathbf{F}_i^* \left(\mathbf{r}_{i,s+1}\right) \cdot \frac{\partial f_{s+1}^*}{\partial \mathbf{p}_i^*} d\mathbf{r}_{s+1}^* d\mathbf{p}_{s+1}^* \tag{6.8}$$

where

$$N_t = t_0/(l_0/v_0) \tag{6.9}$$

$$N_\alpha = \left(\frac{mv_0^2}{\phi_0}\right)\left(\frac{\lambda_0}{l_0}\right) \tag{6.10}$$

and

$$N_\beta = n_0 \lambda_0^3 \tag{6.11}$$

are dimensionless groups.

For most problems, the characteristic velocity would be the molecular velocity and, thus, $mv_0^2 \sim kT_0$, where T_0 is a characteristic temperature. Also, the characteristic intermolecular potential ϕ_0 is usually on the order of magnitude of kT_0, so we can select $\phi_0 = kT_0$ and $mv_0^2 = kT_0$ giving

$$N_\alpha = \frac{\lambda_0}{l_0} \tag{6.12}$$

Specifically in Eq. (6.12), l_0 is the length scale over which an appreciable change in f_s with respect to $\mathbf{r}_i$ occurs and λ_0 is the length scale over which an appreciable change in $\mathbf{F}_i$ with respect to $\mathbf{r}_{ij}$ occurs. For example, in most gas and liquid systems $\lambda_0 << l_0$, and this disparity is utilized to obtain approximate asymptotic solutions to the reduced Liouville equation as discussed more fully hereafter. In general, however, we can distinguish three types of physical systems through the dimensionless groups N_α and N_β (see Fig 6.1), viz.,

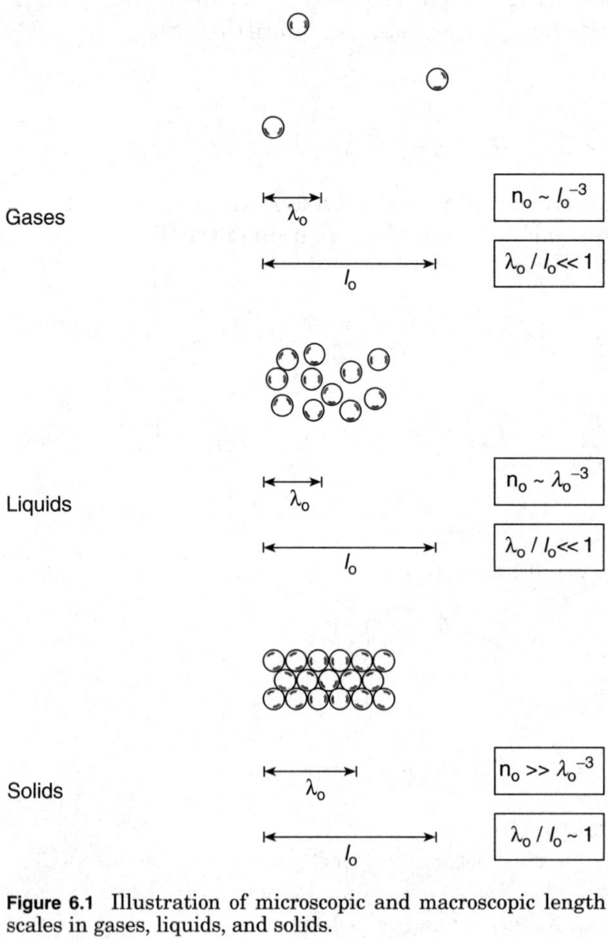

Figure 6.1 Illustration of microscopic and macroscopic length scales in gases, liquids, and solids.

1. Dilute gases ($n_0 \sim l_0^{-3}$):

$$N_\alpha \equiv \frac{\lambda_0}{l_0} << 1 \tag{6.13}$$

and it therefore follows that

$$N_\beta \sim \left(\frac{\lambda_0}{l_0}\right)^3 << 1 \tag{6.14}$$

2. Dense gases and liquids ($n_0 \sim \lambda_0^{-3}$):

$$N_\alpha \equiv \frac{\lambda_0}{l_0} << 1 \tag{6.15}$$

and it therefore follows that

$$N_\beta \sim 1 \tag{6.16}$$

3. Solids ($n_0 >> \lambda_0^{-3}, l_0 \sim \lambda_0$):

$$N_\alpha \equiv \frac{\lambda_0}{l_0} \sim 1 \tag{6.17}$$

and it therefore follows that

$$N_\beta >> 1 \tag{6.18}$$

In gases and liquids the macroscopic length scale l_0 is on the order of microns to millimeters and the microscopic length scale λ_0 is on the order of angstroms to nanometers; thus, N_α in both systems is very small. This should be contrasted with solids where the molecular interaction length scale λ_0 can extend great distances on the order of the macroscopic length scale leading to $N_\alpha \sim 1$ and physically to highly spatially ordered molecular states. Also note that in gases the number density is on the order of the macroscopic length scale to the (-3) power; or in other words, l_0^3 is the average volume per molecule in the gas state. In liquids, on the other hand, the average volume per molecule is around λ_0^3. In solids, at nominal temperatures where the intermolecular or interatomic interaction range λ_0 can extend over many molecules, the average volume per molecule can be much less than λ_0^3.

Interlude 6.1 What Exactly is Scaling? In general, the reduced Liouville equation represents a complex, multidimensional partial differential equation that must be solved subject to the boundary and initial conditions of a particular problem. In many cases, however, a number of terms in this equation are negligible, small, or insignificant and, thus, they can be discarded in the mathematical analysis. Common examples involving a reduction in terms, which may be familiar to the reader, include momentum conservation equation problems where viscous terms may be small compared to convection terms, and energy conservation equation problems where energy conduction may be significant in only a single spatial direction. In general, since we do not know the solution to the conservation equations ahead of time, how are we to determine which terms may be significant or insignificant in any particular problem?

The answer to this question is the subject of *scaling and dimensional analysis*. In general, scaling involves the nondimensionalization of the conservation equations where the *characteristic variables* used for nondimensionalization are selected as their *maximum* values, e.g., the maximum values of velocity, temperature, length, and the like, in a particular problem. Let's see specifically how this method works and why it can often lead to a simplification of partial differential equations.

The first step in scaling involves nondimensionalization of the partial differential equation (PDE). To accomplish this we introduce dimensionless variables (denoted by an asterisk) for every dependent and independent variable of the PDE, such as

$$\mathbf{v}^* = \mathbf{v}/v_0$$

and

$$\mathbf{r}^* = \mathbf{r}/r_0$$

where the characteristic variables v_0 and r_0 are to be selected in a special manner. Specifically, in scaling the characteristic variables such as v_0 and r_0—used to nondimensionalize the PDE—are selected as the maximum values that exist in that particular problem. In doing this, we are attempting to set the dimensionless derivative terms close to one throughout the domain in question as illustrated in Fig. 6.2 The behavior shown in Fig. 6.2 is called "orthodox," meaning commonly observed. However, if the dependent variables

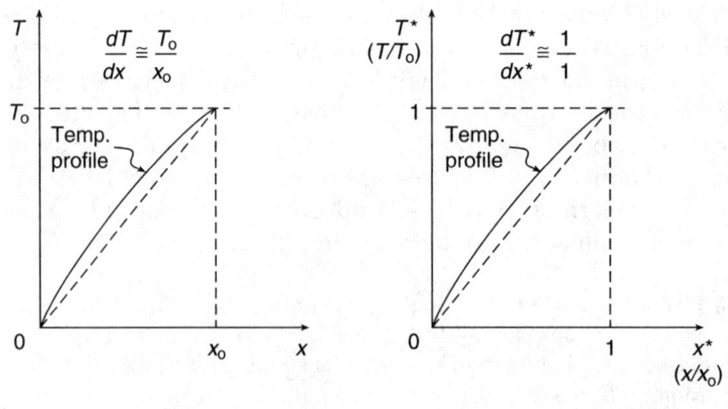

Figure 6.2 Orthodox behavior of a simple derivative.

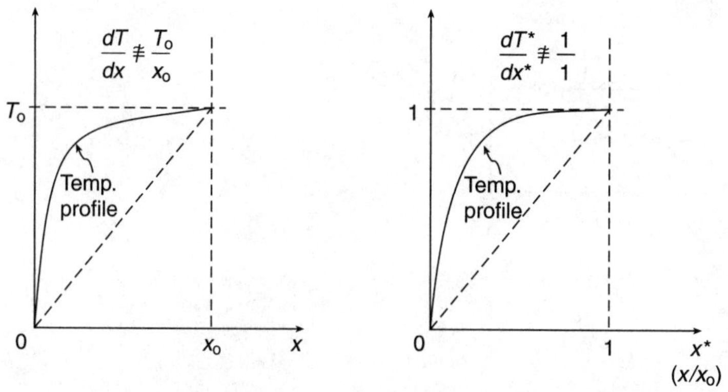

Figure 6.3 Unorthodox behavior of a simple derivative.

(velocity or temperature) vary "dramatically" over the domain (Fig. 6.3), then the dimensionless derivatives will not necessarily be close to one throughout the domain. In this latter case, the behavior of the dependent variable is said to be "unorthodox," and using the maximum values for the characteristic variables does not properly scale the transport equations.†

As a simple example of orthodox behavior, consider the problem of two-dimensional energy conduction through a long solid bar as shown in Fig. 6.4. The temperatures along the boundaries are maintained as shown in Fig. 6.4. With the dimensionless, scaled temperatures defined as $T^* = (T - T_0)/(T_1 - T_0)$, the steady-state conduction equation is

$$\frac{\partial^2 T^*}{\partial x^{*2}} + \left(\frac{x_0}{z_0}\right)^2 \frac{\partial^2 T^*}{\partial z^{*2}} = 0$$

Now with $z_0 >> x_0$ (long bar), the coefficient of the second term (z-conduction term) will always be extremely small. Also, our scaling has resulted in both $\partial^2 T^*/\partial x^{*2}$ and $\partial^2 T^*/\partial z^{*2}$ having approximate values of one or less, except at the ends (Fig. 6.4). Therefore, we can safely neglect the second term compared to the first term, i.e., the x-conduction term, as long as we stay away from the ends. This is the mathematical reasoning behind neglecting conduction along the length of a bar for extremely long bars.

As an example of unorthodox behavior, consider the large velocity (large Reynolds number) flow over a sphere as shown in Fig. 6.5. We might assume that the characteristic length scale in this problem is the sphere diameter (d), that is, the Reynolds number, which is the ratio of fluid inertial to viscous terms, is postulated to be

$$N_{\text{Re}} = \frac{\rho d v_\infty}{\mu}$$

†For further examples and discussion, see Lin and Segel (1974) in the Further Reading section at the end of this chapter.

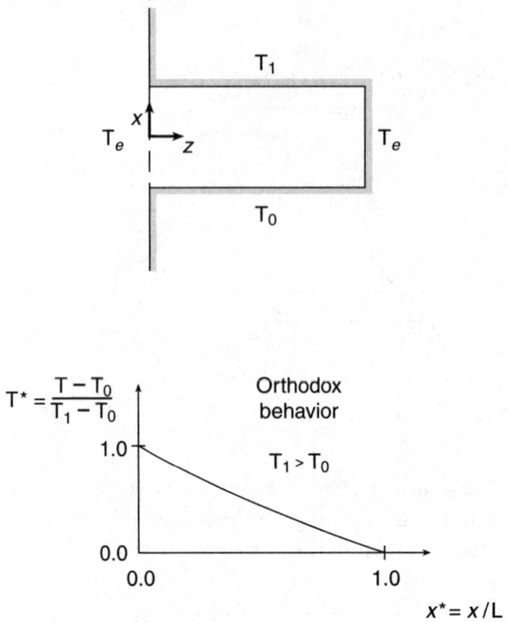

Figure 6.4 Two-dimensional temperature variation problem—example of orthodoxy.

where ρ and μ are the density and viscosity of the fluid, respectively, and v_∞ is the so-called free stream velocity. Thus we are led to the condition $N_{\text{Re}} >> 1$ for large v_∞. According to the scaled transport equation for momentum, developed later in this chapter, the viscous terms should be negligible under this condition. The actual velocity field is shown in Fig. 6.5 and, as can be seen, the velocity changes sharply over a very thin region near the surface known as the momentum boundary layer region. The proper length scale in this region is the thickness of the boundary layer region (l_b), which is known to be much, much less than d. The proper Reynolds number is actually

$$N_{\text{Re}} = \frac{\rho l_b v_\infty}{\mu}$$

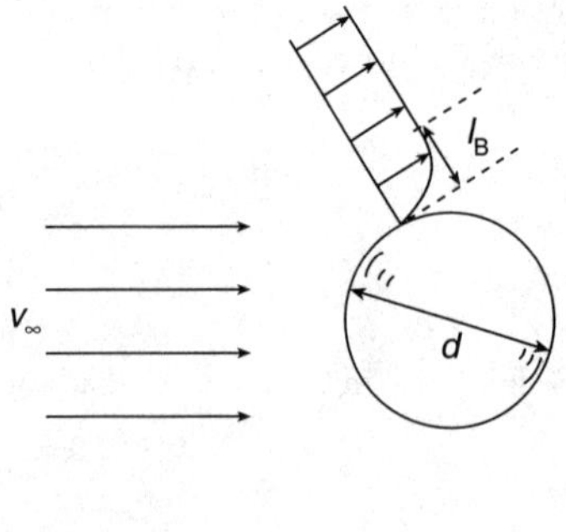

Figure 6.5 Velocity profile near a solid sphere—example of unorthodoxy.

which is typically *not* large even for large v_∞ and, thus, both fluid inertial and viscous terms are important in the boundary layer region. In this problem we can assign different length scales in different parts of the domain and the solutions can be matched together (sort of a mathematical patching or mending).

All of this may seem a little perplexing, but there are formal ways of quantifying the unorthodox behavior in the previous example. The techniques fall under the general subject of *perturbation theory* and are beyond the scope of this text.[†] Fortunately, there are a significant number of important practical problems where orthodox behavior holds and straightforward scaling of the PDE may lead to significant simplifications.

6.3 Regular Perturbative Expansion of the Reduced Liouville Equation for Dense Gases and Liquids

As noted in the previous chapter, to "close" the transport equations an expression for two-molecule density function $f_2(\mathbf{r}, \mathbf{r}', \mathbf{p}, \mathbf{p}', t)$ is needed. Approximate solutions to the reduced Liouville equation for the case $s = 2$ are therefore sought. Of course, if we were able to obtain the complete or exact solution to the function f_2 from the reduced Liouville equation, for any given system in a nonequilibrium state, then the local spatial and temporal thermodynamic state functions could be obtained directly from their basic definitions (at least for pairwise additive systems) and the solution via the transport equations becomes unnecessary or superfluous. Unfortunately, complete or exact solutions to the reduced Liouville equation for nonequilibrium systems are extremely difficult to obtain, even by numerical means, so that the transport route is often our only recourse.

A straightforward approximate approach to obtain the solution to Eq. (6.8) for $s = 2$, is possible under conditions

$$N_\beta = n_0\lambda_0^3 \sim 1 \tag{6.19}$$

$$N_\alpha (= \lambda_0 / l_0) << 1 \tag{6.20}$$

i.e., a dense gas or liquid system where $\lambda_0 \sim n_0^{-3}$ and the intermolecular interaction, microscopic length scale λ_0, is much smaller than the macroscopic length scale l_0. We also choose the characteristic time scale t_0 to be the macroscopic time scale (l_0/v_0), and thus

$$N_t = 1 \tag{6.21}$$

[†]See Kervorkian and Cole (1981), Van Dyke (1975), and Nayfeh (1973) in the Further Reading section at the end of this chapter.

Any time dependency in our resulting solution should be associated with macroscopic variables only. Under the above conditions, the solution is written in the following simple perturbation type expansion

$$f_2^* = f_2^{*(0)} + \varepsilon f^{*(1)} + \varepsilon^2 f_2^{*(2)} + \cdots \tag{6.22}$$

where ε is the so-called "smallness" parameter, $\varepsilon \equiv N_\alpha$, $f_2^{*(0)}$ is called the leading or zero-order solution, $f_2^{*(1)}$ is called the first-order solution, and so on. Now, if the reduced Liouville equation has been scaled properly (see Interlude 6.1), each of the terms involving $f_2^{*(0)}$, $f_2^{*(1)}$, $f_2^{*(2)}$, and so on will have a maximum value ~ 1, and the series solution, Eq. (6.22), will be rapidly convergent for $\varepsilon << 1$. From here on in this section, we will drop the asterisk notation and all quantities will be considered dimensionless unless otherwise noted. Also, without loss of generality, we can express f_2 in terms of the center of mass $\mathbf{r}_{\mathrm{cm}}$ $[\mathbf{r}_{\mathrm{cm}} \equiv (\mathbf{r}_1+\mathbf{r}_2)/2]$ and relative positional coordinate $\mathbf{r}_{12}(\mathbf{r}_{12} \equiv \mathbf{r}_2-\mathbf{r}_1)$ as

$$f_2(\mathbf{r}_1, \mathbf{r}_2, \mathbf{p}_1, \mathbf{p}_2, t) = \tilde{f}_2(\mathbf{r}_{\mathrm{cm}}, \mathbf{r}_{12}, \mathbf{p}_1, \mathbf{p}_2, t) \tag{6.23}$$

where we note that the jacobian of this transformation is unity.[1] Also, note that

$$\frac{\partial f_2(\mathbf{r}_1, \mathbf{r}_2, \mathbf{p}_1, \mathbf{p}_2, t)}{\partial \mathbf{r}_1} = \frac{1}{2}\frac{\partial \tilde{f}_2}{\partial \mathbf{r}_{\mathrm{cm}}} - \frac{1}{\varepsilon}\frac{\partial \tilde{f}_2}{\partial \mathbf{r}_{12}} \tag{6.24}$$

$$\frac{\partial f_2(\mathbf{r}_1, \mathbf{r}_2, \mathbf{p}_1, \mathbf{p}_2, t)}{\partial \mathbf{r}_2} = \frac{1}{2}\frac{\partial \tilde{f}_2}{\partial \mathbf{r}_{\mathrm{cm}}} + \frac{1}{\varepsilon}\frac{\partial \tilde{f}_2}{\partial \mathbf{r}_{12}} \tag{6.25}$$

$$= \frac{1}{2}\frac{\partial \tilde{f}_2}{\partial \mathbf{r}_{\mathrm{cm}}} - \frac{1}{\varepsilon}\frac{\partial \tilde{f}_2}{\partial \mathbf{r}_{21}}$$

where $\mathbf{r}_{21} = \mathbf{r}_1 - \mathbf{r}_2 = -\mathbf{r}_{12}$.

Now, substituting Eqs.(6.22) through (6.25) into Eq. (6.8) written for $s = 2$, and multiplying through by ε, and collecting terms of like order in ε leads to the following problems (Prob. 6.2):

$$\varepsilon^0 : -\mathbf{p}_1 \cdot \frac{\partial \tilde{f}_2^{(0)}}{\partial \mathbf{r}_{12}} - \mathbf{p}_2 \cdot \frac{\partial \tilde{f}_2^{(0)}}{\partial \mathbf{r}_{21}} + \mathbf{F}_1(\mathbf{r}_{12}) \cdot \frac{\partial \tilde{f}_2^{(0)}}{\partial \mathbf{p}_1} + \mathbf{F}_2(\mathbf{r}_{21}) \cdot \frac{\partial \tilde{f}_2^{(0)}}{\partial \mathbf{p}_2}$$

$$= -\int\int \mathbf{F}_1(\mathbf{r}_{13}) \cdot \frac{\partial f_3^{(0)}}{\partial \mathbf{p}_1} d\mathbf{r}_3 d\mathbf{p}_3 - \int\int \mathbf{F}_2(\mathbf{r}_{23}) \cdot \frac{\partial f_3^{(0)}}{\partial \mathbf{p}_2} d\mathbf{r}_3 d\mathbf{p}_3 \tag{6.26}$$

$$\varepsilon^1 : \frac{\partial \tilde{f}_2^{(0)}}{\partial t} + \left(\frac{\mathbf{p}_1 + \mathbf{p}_2}{2}\right) \cdot \frac{\partial \tilde{f}_2^{(0)}}{\partial \mathbf{r}_{\text{cm}}} - \mathbf{p}_1 \cdot \frac{\partial \tilde{f}_2^{(1)}}{\partial \mathbf{r}_{12}} - \mathbf{p}_2 \cdot \frac{\partial \tilde{f}_2^{(1)}}{\partial \mathbf{r}_{21}}$$
$$+ \mathbf{F}_1(\mathbf{r}_{12}) \cdot \frac{\partial \tilde{f}_2^{(1)}}{\partial \mathbf{p}_1} + \mathbf{F}_2(\mathbf{r}_{21}) \cdot \frac{\partial \tilde{f}_2^{(1)}}{\partial \mathbf{p}_2}$$
$$= - \int\!\!\int \mathbf{F}_1(\mathbf{r}_{13}) \cdot \frac{\partial f_3^{(1)}}{\partial \mathbf{p}_1} d\mathbf{r}_3 d\mathbf{p}_3 - \int\!\!\int \mathbf{F}_2(\mathbf{r}_{23}) \cdot \frac{\partial f_3^{(1)}}{\partial \mathbf{p}_2} d\mathbf{r}_3 d\mathbf{p}_3 \tag{6.27}$$

and so on.

Now, we will show that the local equilibrium behavior of f_2 and f_3 [see Chap. 5, Eq. (5.93)] is a particular solution to Eq. (6.26),[†] i.e., in dimensionless terms[‡]

$$\tilde{f}_2^{(0)} = \frac{\tilde{n}_2^{(0)}(\mathbf{r}_{\text{cm}}, \mathbf{r}_{12}, t)}{[2\pi T(\mathbf{r}_{\text{cm}}, t)]^3} \exp\left\{ - \sum_{j=1}^{2} \frac{[\mathbf{p}_j - \mathbf{v}_0(\mathbf{r}_{\text{cm}}, t)]^2}{2T(\mathbf{r}_{\text{cm}}, t)} \right\} \tag{6.28}$$

$$f_3^{(0)} = \frac{n_3^{(0)}}{[2\pi T(\mathbf{r}_{\text{cm}}, t)]^{9/2}} \exp\left\{ - \sum_{j=1}^{3} \frac{[\mathbf{p}_j - \mathbf{v}_0(\mathbf{r}_{\text{cm}}, t)]^2}{2T(\mathbf{r}_{\text{cm}}, t)} \right\} \tag{6.29}$$

where $\tilde{n}_2^{(0)}$, $n_3^{(0)}$, T, $\mathbf{p}_j$, and $\mathbf{v}_0$ are all dimensionless quantities. Note that due to its macroscopic basis we are neglecting any differences between the center of mass position in the two-body and three-body interacting systems as it appears in $\mathbf{v}_0$ and T. To show that Eq. (6.27) and Eq. (6.28) are particular solutions to Eq. (6.26), we follow the same methodology as in Chap. 4 in the development of the equilibrium solution to the Liouville equation. Thus, we write a particular, separable solution as

$$\tilde{f}_2^{(0)} = \tilde{n}_2^{(0)}(\mathbf{r}_{\text{cm}}, \mathbf{r}_{12}, t) h(\mathbf{p}_1) h(\mathbf{p}_2) \tag{6.30}$$

and

$$f_3^{(0)} = n_3^{(0)} h(\mathbf{p}_1) h(\mathbf{p}_2) h(\mathbf{p}_3) \tag{6.31}$$

[cf. Eq. (4.2)]

[†]Of course, we may raise the same issues here of existence and uniqueness that were discussed in Chap. 4 concerning the equilibrium solution to the Liouville equation.

[‡]$T^*(\mathbf{r}, t) = T(\mathbf{r}, t)/T_0, n_2^* = n_2/n_0^6, \ldots$

Substituting Eqs. (6.30) and (6.31) into Eq. (6.26) gives

$$-\mathbf{p}_1 \cdot \frac{\partial \tilde{n}_2^{(0)}}{\partial \mathbf{r}_{12}} + \frac{\partial \ln h(\mathbf{p}_1)}{\partial \mathbf{p}_1} \cdot \left[\tilde{n}_2^{(0)} \mathbf{F}_1(\mathbf{r}_{12}) + \int \mathbf{F}_1(\mathbf{r}_{13}) n_3^{(0)} d\,\mathbf{r}_3\right]$$
$$-\mathbf{p}_2 \cdot \frac{\partial \tilde{n}_2^{(0)}}{\partial \mathbf{r}_{21}} + \frac{\partial \ln h(\mathbf{p}_2)}{\partial \mathbf{p}_2} \cdot \left[\tilde{n}_2^{(0)} \mathbf{F}_2(\mathbf{r}_{21}) + \int \mathbf{F}_2(\mathbf{r}_{23}) n_3^{(0)} d\,\mathbf{r}_3\right] = 0 \tag{6.32}$$

Now, we can add the term

$$\mathbf{v}_0 \cdot \frac{\partial \tilde{n}_2^{(0)}}{\partial \mathbf{r}_{12}} + \mathbf{v}_0 \cdot \frac{\partial \tilde{n}_2^{(0)}}{\partial \mathbf{r}_{21}} = 0 \tag{6.33}$$

to Eq. (6.32), since

$$\frac{\partial \tilde{n}_2}{\partial \mathbf{r}_{12}} = -\frac{\partial \tilde{n}_2}{\partial \mathbf{r}_{21}} \tag{6.34}$$

Subsequently, following the development of the equilibrium solution in Chap. 4, the momentum space solution to Eq. (6.32) can be written as [cf. Eq. (4.11)]

$$\frac{\partial h}{\partial \mathbf{p}_i} = \frac{-h(\mathbf{p}_i - \mathbf{v}_0)}{T} \qquad i = 1, 2 \tag{6.35}$$

and the (dimensionless) configurational integral equation also follows as

$$\frac{1}{T}\frac{\partial \tilde{n}_2^{(0)}}{\partial \mathbf{r}_{12}} + \tilde{n}_2^{(0)} \mathbf{F}_1(\mathbf{r}_{12}) + \int \mathbf{F}_1(\mathbf{r}_{13}) n_3^{(0)} d\,\mathbf{r}_3 = 0 \tag{6.36}$$

or, equivalently, by subscript rotation ($1 \to 2$ and $2 \to 1$)

$$\frac{1}{T}\frac{\partial \tilde{n}_2^{(0)}}{\partial \mathbf{r}_{21}} + \tilde{n}_2^{(0)} \mathbf{F}_2(\mathbf{r}_{21}) + \int \mathbf{F}_2(\mathbf{r}_{23}) n_3^{(0)} d\,\mathbf{r}_3 = 0 \tag{6.37}$$

which, on comparison to Eq. (4.49), proves that the local equilibrium solution is indeed a particular solution to the leading-order equation, Eq. (6.26).

We note that the first-order ($s = 1$) distribution function follows to leading-order from the leading-order $\tilde{f}_2^{(0)}$ solution, Eq. (6.28), as

$$\tilde{f}_1^{(0)}(\mathbf{r}_{\mathrm{cm}}, \mathbf{p}_1, t) = \int \tilde{f}_2^{(0)} d\,\mathbf{r}_{12}\, d\,\mathbf{p}_2$$
$$= \frac{n_1(\mathbf{r}_{\mathrm{cm}}, t)}{[2\pi T(\mathbf{r}_{\mathrm{cm}}, t)]^{3/2}} \exp\left\{-\frac{[\mathbf{p}_1 - \mathbf{v}_0(\mathbf{r}_{\mathrm{cm}}, t)]^2}{2T(\mathbf{r}_{\mathrm{cm}}, t)}\right\} \tag{6.38}$$

We can also see that the expression for $\tilde{f}_2^{(0)}$, Eq. (6.28), is consistent with the leading-order (ε^0) equation for $s = 1$, obtainable from Eq. (6.28) as

$$\varepsilon^0 : 0 = \int\int \mathbf{F}_{12}(\mathbf{r}_{12}) \cdot \frac{\partial \tilde{f}_2^{(0)}}{\partial \mathbf{p}_1} d\mathbf{r}_{12}\, d\mathbf{p}_2 \tag{6.39}$$

where this easily follows by noting that $\tilde{f}_2^{(0)}$ in Eq. (6.28) is symmetric with respect to the sign of $\mathbf{r}_{12}$, whereas $\mathbf{F}_{12}$ is antisymmetric and thus the integral given in Eq. (6.39) vanishes identically. An example of a lack in self-consistency of solution behavior will be seen later in this chapter when dilute gases are examined.

It is interesting to note that Eqs.(6.26) through (6.28) were first given some time ago by Born and Green.[2] Unfortunately, the solution to the first-order (ε^1) equation, Eq. (6.27), has proved to be a formidable task. Let's now examine its behavior in more detail.

First, we will need expressions for $\partial \tilde{f}_2^{(0)}/\partial t$ and $\partial \tilde{f}_2^{(0)}/\partial \mathbf{r}_{\text{cm}}$. To simplify matters somewhat, let's consider a *constant density* system and focus solely on macroscopic velocity ($\mathbf{v}_0$) and temperature gradients (T). By the product rule, we can write

$$\begin{aligned}\frac{\partial \tilde{f}_2^{(0)}}{\partial t} &= \frac{\partial \tilde{f}_2^{(0)}}{\partial T}\frac{\partial T}{\partial t} + \frac{\partial \tilde{f}_2^{(0)}}{\partial \mathbf{v}_0} \cdot \frac{\partial \mathbf{v}_0}{\partial t} \\ &= \tilde{f}_2^{(0)} \left[\frac{1}{T}\left(\left(\sum_{i=1}^{2} \frac{\mathbf{p}_i'^2}{2T} \right) - 3 \right) + \frac{\partial \ln \tilde{n}_2^{(0)}}{\partial T} \right] \frac{\partial T}{\partial t} + \sum_{i=1}^{2} \frac{\mathbf{p}_i'}{T} \tilde{f}_2^{(0)} \cdot \frac{\partial \mathbf{v}_0}{\partial t}\end{aligned} \tag{6.40}$$

and

$$\begin{aligned}\frac{(\mathbf{p}_1 + \mathbf{p}_2)}{2} \cdot \frac{\partial \tilde{f}_2^{(0)}}{\partial \mathbf{r}_{\text{cm}}} &= \tilde{f}_2^{(0)} \left[\frac{1}{T}\left(\left(\sum_{i=1}^{2} \frac{\mathbf{p}_i'^2}{2T} \right) - 3 \right) + \frac{\partial \ln \tilde{n}_2^{(0)}}{\partial T} \right] \\ &\times \frac{(\mathbf{p}_1 + \mathbf{p}_2)}{2} \cdot \frac{\partial T}{\partial \mathbf{r}_{\text{cm}}} + \tilde{f}_2^{(0)} \sum_{i=1}^{2} \frac{\mathbf{p}_i'}{T} \cdot \left(\frac{(\mathbf{p}_1 + \mathbf{p}_2)}{2} \cdot \frac{\partial}{\partial \mathbf{r}_{\text{cm}}} \right) \mathbf{v}_0\end{aligned} \tag{6.41}$$

where $\mathbf{p}_i' = \mathbf{p}_i - \mathbf{v}_0$ is the (dimensionless) relative molecular momentum. Now, to this order, we can use the dimensionless transport equations for equilibrium flows, Eqs. (5.100) and (5.101), to express the time derivatives as

$$\frac{\partial T}{\partial t} = -\mathbf{v}_0 \cdot \frac{\partial T}{\partial \mathbf{r}_{\text{cm}}} \tag{6.42}$$

and

$$\frac{\partial \mathbf{v}_0}{\partial t} = -\left(\mathbf{v}_0 \cdot \frac{\partial}{\partial \mathbf{r}_{cm}}\right)\mathbf{v}_0 - \frac{1}{n_1}\frac{1}{\partial \mathbf{r}_{cm}} \cdot p_{eq}\mathbf{I} \tag{6.43}$$

Substituting for the time derivatives in Eq. (6.40), combining with Eq. (6.41), and simplifying, we obtain

$$\begin{aligned}
&\frac{\partial \tilde{f}_2^{(0)}}{\partial t} + \frac{(\mathbf{p}_1 + \mathbf{p}_2)}{2} \cdot \frac{\partial \tilde{f}_2^{(0)}}{\partial \mathbf{r}_{cm}} \\
&= \frac{1}{2}\tilde{f}_2^{(0)} \sum_{i=1}^{2} \mathbf{p}_i' \cdot \frac{\partial T}{\partial \mathbf{r}_{cm}} \left[\frac{1}{T}\left(\left(\sum_{i=1}^{2} \frac{\mathbf{p}_i'^2}{2T}\right) - 3\right) + \frac{\partial \ln \tilde{n}_2^{(0)}}{\partial T} - \frac{2}{n_1}\frac{\partial p_{eq}}{\partial T}\right] \\
&+ \frac{\tilde{f}_2^{(0)}}{2T}\left(\sum_{i=1}^{2} \mathbf{p}_i'\right)\left(\sum_{i=1}^{2} \mathbf{p}_i'\right) : \mathbf{S}
\end{aligned} \tag{6.44}$$

where $\mathbf{S}$ is the so-called rate-of-strain tensor

$$\mathbf{S} \equiv \frac{1}{2}\left[\frac{\partial \mathbf{v}_0}{\partial \mathbf{r}_{cm}} + \left(\frac{\partial \mathbf{v}_0}{\partial \mathbf{r}_{cm}}\right)^t\right] - \frac{1}{3}\left(\frac{\partial}{\partial \mathbf{r}_{cm}} \cdot \mathbf{v}_0\right) \tag{6.45}$$

and the superscript t denotes the transpose operation. Now, we can substitute Eq. (6.44) into Eq. (6.27) and write a solution as a linear combination of each of the macroscopic gradients appearing in the non-homogeneous term given by Eq. (6.44). To further simplify matters, and to illustrate the procedures involved without undue mathematical complexities, we will consider only the case of a macroscopic velocity gradient and assume that the temperature is constant. This assumption can easily be relaxed, albeit at the expense of carrying around more terms. The original Born and Green study[2] can be consulted for the more general case.

Substituting Eq. (6.44) into Eq. (6.27) gives, for the macroscopic velocity gradient problem

$$\begin{aligned}
&-\mathbf{p}_1 \cdot \frac{\partial \tilde{f}_2^{(1)}}{\partial \mathbf{r}_{12}} - \mathbf{p}_2 \cdot \frac{\partial \tilde{f}_2^{(1)}}{\partial \mathbf{r}_{21}} + \mathbf{F}_1(\mathbf{r}_{12}) \cdot \frac{\partial \tilde{f}_2^{(1)}}{\partial \mathbf{p}_1} + \mathbf{F}_2(\mathbf{r}_{21}) \cdot \frac{\partial \tilde{f}_2^{(1)}}{\partial \mathbf{p}_2} \\
&+ \iint \mathbf{F}_1(\mathbf{r}_{13}) \cdot \frac{\partial \tilde{f}_3^{(1)}}{\partial \mathbf{p}_1} d\mathbf{r}_3 d\mathbf{p}_3 + \iint \mathbf{F}_2(\mathbf{r}_{23}) \cdot \frac{\partial \tilde{f}_3^{(1)}}{\partial \mathbf{p}_2} d\mathbf{r}_3 d\mathbf{p}_3 \\
&= \frac{\tilde{f}_2^{(0)}}{2T}\left(\sum_{i=1}^{2} \mathbf{p}_i'\right)\left(\sum_{i=1}^{2} \mathbf{p}_i'\right) : \mathbf{S}
\end{aligned} \tag{6.46}$$

Now it can be seen that the entire macroscopic dependence of the first-order (ε^1) equation, Eq. (6.48), is due to $\mathbf{S}$. Thus, the general form of the solution must be

$$\tilde{f}_2^{(1)} = \frac{1}{T}\tilde{f}_2^{(0)}\mathbf{h}_2^{(1)}(\mathbf{r}_{12}, \mathbf{p}_1, \mathbf{p}_2) : \mathbf{S} \tag{6.47}$$

$$f_3^{(1)} = \frac{1}{T} f_3^{(0)}\mathbf{h}_3^{(1)} : \mathbf{S} \tag{6.48}$$

where the dyadic functions $\mathbf{h}_2^{(1)}$ and $\mathbf{h}_3^{(1)}$ are to be determined.

As noted by Born and Green,[2] a superposition approximation for the three-body distribution function, similar to Kirkwood's form in equilibrium systems, Eq. (4.57), is needed for further analysis of Eq. (6.48). Using the solution form Eq. (6.47), we can, however, immediately write down expressions for the so-called "auxiliary" conditions and the property flux vectors. First, we have to $O(\varepsilon^2)$

$$\tilde{f}_2 = \tilde{f}_2^{(0)} + \varepsilon\frac{\tilde{f}_2^{(0)}}{T}\mathbf{h}_2^{(1)}(\mathbf{r}_{12}, \mathbf{p}_1, \mathbf{p}_2) : \mathbf{S} \tag{6.49}$$

and, thus, from Chap. 3 we can obtain expressions for the required reduced distribution functions

$$\begin{aligned} f_1(\mathbf{r}_{\text{cm}}, \mathbf{p}_1, t) &= \frac{1}{(N-1)}\int \tilde{f}_2\, d\mathbf{r}_{12}\, d\mathbf{p}_2 \\ &= f_1^{(0)} + \frac{\varepsilon}{(N-1)}\frac{1}{T}\mathbf{S} : \int \mathbf{h}_2^{(1)}\tilde{f}_2^{(0)} d\mathbf{r}_{12}\, d\mathbf{p}_2 \end{aligned} \tag{6.50}$$

and

$$\begin{aligned} \tilde{n}_2(\mathbf{r}_{12}, \mathbf{r}_{\text{cm}}, t) &= \int \tilde{f}_2\, d\mathbf{p}_1 d\mathbf{p}_2 \\ &= \tilde{n}_2^{(0)} + \varepsilon\frac{1}{T}\mathbf{S} : \int \mathbf{h}_2^{(1)}\tilde{f}_2^{(0)} d\mathbf{p}_1 d\mathbf{p}_2 \end{aligned} \tag{6.51}$$

Now, in dimensionless forms, we know from the previous chapter that

$$\int f_1 d\mathbf{p} = n \tag{6.52}$$

$$\int \mathbf{p} f_1 d\mathbf{p} = n\mathbf{v}_0 \tag{6.53}$$

$$\frac{1}{2}\int p'^2 f_1 d\mathbf{p} = \frac{3}{2}nT \tag{6.54}$$

Substitution of Eq. (6.50) into Eqs. (6.52) through (6.54) leads to

$$\int f_1^{(0)} d\mathbf{p} = n(\mathbf{r}_{\mathrm{cm}}, t) \tag{6.55}$$

$$\int \mathbf{p} f_1^{(0)} d\mathbf{p} = n(\mathbf{r}_{\mathrm{cm}}, t)\mathbf{v}_0(\mathbf{r}_{\mathrm{cm}}, t) \tag{6.56}$$

$$\frac{1}{2}\int p'^2 f_1^{(0)} d\mathbf{p} = \frac{3}{2} n(\mathbf{r}_{\mathrm{cm}}, t)T \tag{6.57}$$

and the auxiliary conditions

$$\int \mathbf{S} : \mathbf{h}_2^{(1)} \tilde{f}_2^{(0)} d\mathbf{r}_{12}\, d\mathbf{p}_1 d\mathbf{p}_2 = 0 \tag{6.58}$$

$$\int \mathbf{p}_i \mathbf{S} : \mathbf{h}_2^{(1)} \tilde{f}_2^{(0)} d\mathbf{r}_{12}\, d\mathbf{p}_1 d\mathbf{p}_2 = 0 \qquad i = 1, 2 \tag{6.59}$$

$$\int p_i'^2 \mathbf{S} : \mathbf{h}_2^{(1)} \tilde{f}_2^{(0)} d\mathbf{r}_{12}\, d\mathbf{p}_1 d\mathbf{p}_2 = 0 \qquad i = 1, 2 \tag{6.60}$$

We recall from Chap. 5 that the momentum flux tensor consists of both a kinetic and a potential component given in dimensionless form as†

$$\mathbf{P} = \mathbf{P}_k + \mathbf{P}_\Phi \tag{6.61}$$

where

$$\mathbf{P}_k = \int \mathbf{p}_1' \mathbf{p}_1' f_1 d\mathbf{p}_1 \tag{6.62}$$

and

$$\mathbf{P}_\phi = -\frac{1}{2}\int \tilde{n}_2 \frac{\mathbf{R}\mathbf{R}}{R}\frac{d\,\Phi(R)}{dR} d\mathbf{R} \tag{6.63}$$

where $\mathbf{R} \equiv \mathbf{r}_{12}$. Substituting Eq. (6.50) into Eq. (6.62) gives, for the kinetic component,

$$\mathbf{P}_k = \mathbf{P}_k^{(0)} + \varepsilon \mathbf{P}_k^{(1)} \tag{6.64}$$

where

$$\mathbf{P}_k^{(0)} = \int \mathbf{p}_1' \mathbf{p}_1' f_1^{(0)} d\mathbf{p}_1 = n_1 T\,\mathbf{I} \tag{6.65}$$

†Specifically, $\mathbf{P}_k^* = \mathbf{P}_k / P_{k_o}$ and $\mathbf{P}_\Phi^* = \mathbf{P}_\Phi / P_{\Phi_o}$, where the reference values are $P_{k_o} = mn_0 v_0^2 = mv_0^2/(\varepsilon^3 l_0^3)$ and $P_{\phi_0} = n_0^2 \phi_0 l_0^3 = mn_0^2 v_0^2 l_0^3 = mv_0^2/\varepsilon^6 l_0^3$. These simple scaling arguments show that in the liquid state the potential contribution is expected to be three orders of magnitude larger than the kinetic contribution.

and

$$\mathbf{P}_k^{(1)} = \frac{1}{(N-1)}\frac{1}{T}\int\int \mathbf{p}_1'\mathbf{p}_1'\mathbf{S} : \mathbf{h}_2^{(1)}\tilde{f}_2^{(0)} d\mathbf{r}_{12}\, d\mathbf{p}_1 d\mathbf{p}_2 \tag{6.66}$$

Equation (6.63) is simply the dimensionless ideal gas equation of state. For the potential contribution, we substitute Eq. (6.51) into Eq. (6.63) to obtain

$$\mathbf{P}_\phi = \mathbf{P}_\phi^{(0)} + \varepsilon \mathbf{P}_\phi^{(1)} \tag{6.67}$$

where

$$\mathbf{P}_\phi^{(0)} = -\frac{1}{2}\int \tilde{n}_2^{(0)} \frac{\mathbf{RR}}{R}\frac{d\phi}{dR} d\mathbf{R} \tag{6.68}$$

and

$$\mathbf{P}_\phi^{(1)} = -\frac{1}{2}\frac{1}{T}\int \mathbf{S} : \mathbf{h}_2^{(1)}\tilde{f}_2^{(0)} \frac{\mathbf{RR}}{R}\frac{d\phi}{dR} d\mathbf{R}\, d\mathbf{p}_1 d\mathbf{p}_2 \tag{6.69}$$

Equation (6.68) is the potential part of the equation of state for the fluid at (local) equilibrium. Both first-order corrections to the momentum flux tensor, Eqs. (6.66) and (6.69), are proportional to the rate-of-strain tensor. The specific form of $\mathbf{h}_2^{(0)}$ therefore expresses the particular rheological behavior of the fluid. Born and Green[2] also considered the problem of simultaneous macroscopic velocity, temperature, and density gradients. All cases, however, lead to integro-differential type equations for the first-order corrections to the local equilibrium state. The general difficulty in obtaining the solution to integro-differential equations, such as Eq. (6.46), has led to alternative and simplified model approaches to the kinetic theory of liquids, which will not be considered here (see the Further Reading section at the end of this chapter). Nonequilibrium liquid state molecular theory still represents a rather open and lucrative area for research, and the Born and Green method outlined here represents a rigorous approach.[2] In order to illustrate the complete determination of the flux expressions, however, including entropy flux and entropy generation, we now turn to the more developed problem of the molecular theory of gases.

6.4 Perturbation Expansion for Dilute Gases

Now let's consider the solution to Eq. (6.8) for gases. We will require $s = 1$ and $s = 2$ to evaluate the property flux terms. Now, as already noted, for gases $N_\alpha = \lambda_0/l_0 << 1$ and $N_\beta = n_0\lambda^3 << 1$. We will also select $N_t = 1$ or $t_0 = l_0/v_0$, the macroscopic time scale and, as in Sec. 6.3, any time dependency of our solution is assumed to be associated with the macroscopic variables only. Later on in our development we will have to question this now embattled assumption. The ratio N_β/N_α in Eq. (6.8) is often expressed in terms of the gas mean-free path from elementary kinetic theory as follows. First, we have

$$\frac{N_\beta}{N_\alpha} = \frac{n_0\lambda_0^3}{\lambda_0/l_0} = n_0\lambda_0^2 l_0 \tag{6.70}$$

and from elementary kinetic theory[†] the reference number density is approximately

$$n_0 \sim \left(\lambda_0^2 \lambda_{mfp}\right)^{-1} \tag{6.71}$$

where λ_{mfp} is the gas mean-free path, i.e., the average distance between molecular collisions. Thus

$$\frac{N_\beta}{N_\alpha} \sim \frac{l_0}{\lambda_{mfp}} \equiv \frac{1}{N_{Kn}} \tag{6.72}$$

where $N_{Kn} \equiv \lambda_{mfp}/l_0$ is called the Knudsen number in gas kinetic theory; for the time being we will consider it to be ~ 1.

Equation (6.8) can now be written for $s = 1$ and $s = 2$ as, respectively

$$\frac{\partial f_1}{\partial t} + \mathbf{p}_1 \cdot \frac{\partial f_1}{\partial \mathbf{r}_1} = -\frac{1}{N_{Kn}} \int\!\!\int \mathbf{F}_1(\mathbf{r}_{12}) \cdot \frac{\partial f_2}{\partial \mathbf{p}_1}\, d\mathbf{r}_2\, d\mathbf{p}_2 \tag{6.73}$$

and

$$\frac{\partial f_2}{\partial t} + \sum_{i=1}^{2} \mathbf{p}_i \cdot \frac{\partial f_2}{\partial \mathbf{r}_i} + \frac{1}{\varepsilon}\left[\mathbf{F}_1(\mathbf{r}_{12}) \cdot \frac{\partial f_2}{\partial \mathbf{p}_1} + \mathbf{F}_2(\mathbf{r}_{21}) \cdot \frac{\partial f_2}{\partial \mathbf{p}_2}\right]$$
$$= -\frac{1}{N_{Kn}}\left[\int\!\!\int \mathbf{F}_1(\mathbf{r}_{13}) \cdot \frac{\partial f_3}{\partial \mathbf{p}_1} d\mathbf{r}_3 d\mathbf{p}_3 + \int\!\!\int \mathbf{F}_2(\mathbf{r}_{23}) \cdot \frac{\partial f_3}{\partial \mathbf{p}_2} d\mathbf{r}_3 d\mathbf{p}_3\right] \tag{6.74}$$

where the smallness parameter $\varepsilon = N_\alpha$.

Now, let's first look at the equation for f_2, Eq. (6.79). As done previously for dense gases and liquids, multiply through Eq. (6.74) by ε, substitute Eqs. (6.24) and (6.25) for the center-of-mass relative coordinates, then substitute the expansions

$$\tilde{f}_2 = \tilde{f}_2^{(0)} + \varepsilon \tilde{f}_2^{(1)} + \cdots \tag{6.75}$$

$$f_3 = f_3^{(0)} + \varepsilon f_3^{(1)} + \cdots \tag{6.76}$$

and finally collect terms of like order in ε to obtain

[†]See the Further Reading section at the end of Chap. 3.

$(s=2)\varepsilon^0$:

$$-\mathbf{p}_1 \cdot \frac{\partial \tilde{f}_2^{(0)}}{\partial \mathbf{r}_{12}} - \mathbf{p}_2 \cdot \frac{\partial \tilde{f}_2^{(0)}}{\partial \mathbf{r}_{21}} + \mathbf{F}_1(\mathbf{r}_{12}) \cdot \frac{\partial \tilde{f}_2^{(0)}}{\partial \mathbf{p}_1} + \mathbf{F}_2(\mathbf{r}_{21}) \cdot \frac{\partial \tilde{f}_2^{(0)}}{\partial \mathbf{p}_2} = 0 \tag{6.77}$$

$(s=2)\varepsilon^1$:

$$\begin{aligned} &\frac{\partial \tilde{f}_2^{(0)}}{\partial t} + \frac{(\mathbf{p}_1+\mathbf{p}_2)}{2} \cdot \frac{\partial \tilde{f}_2^{(0)}}{\partial \mathbf{r}_{cm}} + \frac{1}{N_{Kn}} \\ &\times \left[\int\!\!\int \mathbf{F}_1(\mathbf{r}_{13}) \cdot \frac{\partial f_3^{(0)}}{\partial \mathbf{p}_1} d\mathbf{r}_3 d\mathbf{p}_3 + \int\!\!\int \mathbf{F}_{23}(\mathbf{r}_{23}) \cdot \frac{\partial f_3^{(0)}}{\partial \mathbf{p}_2} d\mathbf{r}_3 d\mathbf{p}_3 \right] \\ &= \mathbf{p}_1 \cdot \frac{\partial \tilde{f}_2^{(1)}}{\partial \mathbf{r}_{12}} + \mathbf{p}_2 \cdot \frac{\partial \tilde{f}_2^{(1)}}{\partial \mathbf{r}_{21}} - \mathbf{F}_1(\mathbf{r}_{12}) \cdot \frac{\partial \tilde{f}_2^{(1)}}{\partial \mathbf{p}_1} - \mathbf{F}_2(\mathbf{r}_{21}) \cdot \frac{\partial \tilde{f}_2^{(1)}}{\partial \mathbf{p}_2} \end{aligned} \tag{6.78}$$

Following the same expansion method for $s = 1, (f_1 = f_1^{(0)} + \varepsilon f_1^{(1)} + \cdots)$, we also obtain

$(s=1)\varepsilon^0$:

$$\frac{\partial f_1^{(0)}}{\partial t} + \mathbf{p}_1 \cdot \frac{\partial f_1^{(0)}}{\partial \mathbf{r}_1} = -\frac{1}{N_{Kn}} \int\!\!\int \mathbf{F}_1(\mathbf{r}_{12}) \cdot \frac{\partial f_2^{(0)}}{\partial \mathbf{p}_1} d\mathbf{r}_2\, d\mathbf{p}_2 \tag{6.79}$$

$(s=1)\varepsilon^1$:

$$\frac{\partial f_1^{(1)}}{\partial t} + \mathbf{p}_1 \cdot \frac{\partial f_1^{(1)}}{\partial \mathbf{r}_1} = -\frac{1}{N_{Kn}} \int\!\!\int \mathbf{F}_1(\mathbf{r}_{12}) \cdot \frac{\partial f_2^{(1)}}{\partial \mathbf{p}_1} d\mathbf{r}_2\, d\mathbf{p}_2 \tag{6.80}$$

Now, it can be seen that Eq. (6.77), describing the leading-order, two-body distribution function, is identical to Eq. (3.29), previously given in Chap. 3 in the development of Boltzmann's collision integral term. Furthermore, the general form of Boltzmann's collision integral is given by the right-hand side of Eq. (6.79). Thus, the leading-order behavior of f_1, Eq. (6.79), precisely follows the Boltzmann transport equation, Eq. (3.37), under the same assumptions following Eq. (3.29) in Chap. 3.

It is also interesting to note that although the local, two-body equilibrium solution is indeed a particular solution to Eq. (6.77), substitution of the equilibrium solution into Eq. (6.79) leads to nonsensical results. Therefore, the equilibrium solution is not self-consistent to leading order. There are numerous extensive treatments on the development, modifications, and treatment of the Boltzmann transport equation in gas kinetic theory (see the Further Reading section at the end of Chap. 3). More modern approaches, originally due to Bogolubov,[3] consider the time scales of the expansion procedure more carefully. These so-called *multiple time-scales* methods are powerful procedures to obtain the correct asymptotic analysis of gas kinetic equations. Without going into details of the expansion here, in short we have incorrectly assumed that the only time dependency in our solution is associated with macroscopic variables. Although this assumption seems to hold well for dense gases and liquids, in dilute gases, where molecular interactions are relatively less frequent, the so-called *microscopic* fluid relaxation times are not small and cannot be ignored in the time-scaling analysis. We also note that the closure results of multiple time-scale approaches, in general, are cast in terms of time correlation functions, which can be readily evaluated by modern numerical computations. (See the Further Reading section at the end of this chapter.)

In order to illustrate the resolution of the property flux terms for gases, which is our immediate focus here, we will turn to the classic, Chapman-Enskog solution to the Boltzmann transport equation with the reservations, discussed earlier, in mind. In the Chapman-Enskog approach, the solution is further expanded in powers of the Knudsen number, assuming that the Knudsen number is small in some sense. Of course, the problem with this particular hypothesis, which is evident from Eqs. (6.73) and (6.74), is that small Knudsen numbers, say $0(\varepsilon)$, require that three-body effects be included; and we are simply back to the previous problem of dense gases and liquids. So, in the Chapman-Enskog expansion method the Knudsen number should be small, say $0(\varepsilon^{1/2})$, but not too small, say $0(\varepsilon)$. Despite all of the shortcomings noted here, results from the Boltzmann transport equation and Chapman-Enskog solution method have been shown to compare well to experimental measurements in gases; and they have provided great insight into the specific closure forms of the transport equations. We will therefore present the method in the next section at least for the purpose of illustrating the fine details of closure methods.

6.5 Chapman-Enskog Solution to the Boltzmann Transport Equation

In dimensionless form, the Boltzmann transport equation, Eq. (3.37), including the possibility of an external force, can be written as

$$\begin{aligned}
&\frac{\partial f_1}{\partial t} + \mathbf{p}_1 \cdot \frac{\partial f_1}{\partial \mathbf{r}_1} + N_F \mathbf{F}_{\text{ext}}(\mathbf{r}_1) \cdot \frac{\partial f_1}{\partial \mathbf{p}_1} \\
&\quad = \frac{1}{N_{Kn}} \int \int_0^{2\pi} \int_0^{b_0} v_{12} \left[f_1(\mathbf{r}_1', \mathbf{p}_1', t) f_1(\mathbf{r}_2', \mathbf{p}_2', t) \right. \\
&\quad \left. - f_1(\mathbf{r}_1, \mathbf{p}_1, t) f_1(\mathbf{r}_2, \mathbf{p}_2, t) \right] b\, db\, d\varepsilon\, d\mathbf{p}_{12}
\end{aligned} \tag{6.81}$$

where $N_F \equiv F_{\text{ext}_0} m/p_0^2$ and it is understood that f_1 represents the leading-order (ε) expansion, i.e., we drop the superscript (0) on f_1 for the sake of simplicity of notation.

As mentioned previously, the Chapman-Enskog solution is based on a small Knudsen number expansion of the Boltzmann equation. Under the conditions of a small Knudsen number, we substitute the following expansion

$$f_1 = f_{(0)} + N_{Kn} f_{(1)} + N_{Kn}^2 f_{(2)} + \cdots \tag{6.82}$$

into Eq. (6.81), which leads to the following problems in order of the Knudsen number: (zero order) N_{Kn}^0:

$$\int\int_0^{2\pi}\int_0^{b_0} v_{21}\left(f'_{(0)_1} f'_{(0)_2} - f_{(0)_1} f_{(0)_2}\right) b\,db\,d\varepsilon\,d\mathbf{p}_{12} = 0 \tag{6.83}$$

(first order) N_{Kn}^1:

$$\begin{aligned}
&\frac{\partial f_{(0)_1}}{\partial t} + \mathbf{p}_1 \cdot \frac{\partial f_{(0)_1}}{\partial \mathbf{r}_1} + N_F \mathbf{F}_{1\text{ext}} \cdot \frac{\partial f_{(0)_1}}{\partial \mathbf{p}_1} \\
&\quad = \int\int_0^{2\pi}\int_0^{b_0} v_{12}\left(f'_{(0)_1} f'_{(1)_2} - f_{(0)_1} f_{(1)_2}\right) b\,db\,d\varepsilon\,d\mathbf{p}_{12} \\
&\qquad + \int\int_0^{2\pi}\int_0^{b_0} v_{12}\left(f'_{(1)_1} f'_{(0)_2} - f_{(1)_1} f_{(0)_2}\right) b\,db\,d\varepsilon\,d\mathbf{p}_2
\end{aligned} \tag{6.84}$$

and so on, for a higher power of the Knudsen number, where we have used the notation $f'_{(0)_1} = f_{(0)}(\mathbf{r}'_1, \mathbf{p}'_1, t)$, $f'_{(0)_2} = f_{(0)}(\mathbf{r}'_2, \mathbf{p}'_2, t), \ldots$

The solution to Eq. (6.83) follows from Boltzmann's H-theorem (Chap. 3), written in *dimensionless* form as

$$f_{(0)_1} = \frac{n_1(\mathbf{r}_1, t)}{(2\pi T)^{3/2}} \exp\left[-\frac{(\mathbf{p}_1 - \mathbf{v}_0)^2}{2T}\right] \tag{6.85}$$

and

$$f_{(0)_2} = \frac{n_1(\mathbf{r}_2, t)}{(2\pi T)^{3/2}} \exp\left[-\frac{(\mathbf{p}_2 - \mathbf{v}_0)^2}{2T}\right] \tag{6.86}$$

Equations (6.85) and (6.86) reflect the local equilibrium nature of the solution to Eq. (6.83). It is, therefore, seen that the so-called Chapman-Enskog method of the solution is based on an expansion about local equilibrium conditions. Pitfalls of this approach have been previously noted. Writing, without loss of generality,

$$f_{(1)} = f_{(0)}\phi(\mathbf{r}, \mathbf{p}, t) \tag{6.87}$$

and substituting Eqs.(6.85) and (6.86) into Eq. (6.84) and assuming no external force, gives after some manipulations (Prob. 6.3).

$$f_{(0)_1}\left[\left(\mathbf{b}_1 : \frac{\partial}{\partial \mathbf{r}_1}\mathbf{v}_0\right) - \left(\frac{5}{2} - w_1^2\right)\left((\mathbf{p}_1 - \mathbf{v}_0)\cdot\frac{\partial \ln T}{\partial \mathbf{r}_1}\right)\right]$$
$$= \int\int\int f_{(0)_1} f_{(0)_2}\left(\phi_1' + \phi_2' - \phi_1 - \phi_2\right) v_{12} b\, d b\, d\epsilon\, d\mathbf{p}_{12} \quad (6.88)$$

where

$$\mathbf{b}_1 = 2\left[\mathbf{w}_1\mathbf{w}_1 - \frac{1}{3}w_1^2\mathbf{I}\right] \quad (6.89)$$

$$\mathbf{w}_1 = \frac{(\mathbf{p}_1 - \mathbf{v}_0)}{\sqrt{2T}} \quad (6.90)$$

From Eq. (6.88) it can be seen that the function ϕ depends on space only through the spatial derivatives of $\mathbf{v}_0$ and T . Consequently, a general form for the function ϕ is

$$\phi_i = -\mathbf{B}_i : \frac{\partial}{\partial \mathbf{r}_1}\mathbf{v}_0 - \mathbf{A}_i \cdot \frac{\partial \ln T}{\partial \mathbf{r}_1} \quad (6.91)$$

where the functions $\mathbf{B}_i$ and $\mathbf{A}_i$ are yet to be determined. Also, from Eq. (6.89), it can be seen that the general forms for the tensor $\mathbf{B}_i$ and the vector $\mathbf{A}_i$ are, respectively,

$$\mathbf{B}_i = \left[\mathbf{w}_i\mathbf{w}_i - \frac{1}{3}w_i^2\mathbf{I}\right] B(w_i) \quad (6.92)$$

and

$$\mathbf{A}_i = \mathbf{w}_i A(w_i) \quad (6.93)$$

Substituting Eq. (6.92) into Eq. (6.88) leads to the following integral equations for the functions $\mathbf{B}_i$ and $\mathbf{A}_i$

$$f_{(0)_1}\mathbf{b}_1 = -\int\int\int \left(\mathbf{B}_1' + \mathbf{B}_2' - \mathbf{B}_1 - \mathbf{B}_2\right) f_{(0)_1} f_{(0)_2} v_{12} bd\, bd\,\epsilon d\,\mathbf{p}_{12} \quad (6.94)$$

and

$$f_{(0)_1}\left(\frac{5}{2} - w_1^2\right)(\mathbf{p}_1 - \mathbf{v}_0) = \int\int\int \left(\mathbf{A}_1' + \mathbf{A}_2' - \mathbf{A}_1 - \mathbf{A}_2\right)$$
$$\times f_{(0)_1} f_{(0)_2} v_{12} bd\, bd\,\epsilon d\,\mathbf{p}_{12} \quad (6.95)$$

In view of the previously given definitions of the average number density, mass average velocity, and kinetic energy, the function ϕ must satisfy the auxiliary conditions [see Eqs. (6.52) through (6.60)]

$$\int f_{(0)_i}\phi_i d\,\mathbf{p}_i = 0 \qquad i = 1, 2 \tag{6.96}$$

$$\int \mathbf{p}_i f_{(0)_i}\phi_i d\,\mathbf{p}_i = 0 \tag{6.97}$$

$$\int (\mathbf{p}_i - \mathbf{v}_0)^2 f_{(0)_i}\phi_i d\,\mathbf{p}_i = 0 \tag{6.98}$$

and, consequently, from Eq. (6.91)

$$\int (\mathbf{A}_i - \mathbf{w}_i) f_{(0)_i} d\,\mathbf{p}_i = 0 \tag{6.99}$$

Note that the form of the function $\mathbf{B}_i$ is such that Eqs. (6.96) through (6.98) are automatically satisfied.

The solution to the linear integral equations, Eqs. (6.94) and (6.95), are given in detail in various references,[4,5] which enables an accurate determination of the functions $\mathbf{A}_i$ and $\mathbf{B}_i$. Since the solution to linear integral equations is a fairly standard mathematical exercise, we instead turn our attention to the specific forms of the property flux expressions and the final "closed" forms of the conservation equations.

6.6 Property Flux Expressions for Gases

The transport equations associated with the Chapman-Enskog expansion can be obtained from the general equations change given earlier. From the truncated expansion

$$f_{(1)} = f_{(0)}(1 + \phi) \tag{6.100}$$

where

$$\phi = -\mathbf{B} : \frac{\partial \mathbf{v}_0}{\partial \mathbf{r}} - \mathbf{A} \cdot \frac{\partial \ln T}{\partial \mathbf{r}} \tag{6.101}$$

and

$$f_{(0)} = \frac{n(\mathbf{r}, t)}{(2\pi T)^{3/2}} \exp\left[-\frac{(\mathbf{p} - \mathbf{v}_0)^2}{2T}\right] \tag{6.102}$$

the kinetic contribution to the pressure tensor is obtained from its definition, in dimensionless terms, Eq. (6.64),

$$\begin{aligned}
\mathbf{P}_k(\mathbf{r}, t) &= \int (\mathbf{p} - \mathbf{v}_0)(\mathbf{p} - \mathbf{v}_0)\, f_1(\mathbf{r}, \mathbf{p}, t)\, d\,\mathbf{p} \\
&= \int (\mathbf{p} - \mathbf{v}_0)(\mathbf{p} - \mathbf{v}_0)\, f_{(0)}(1 + \phi)\, d\,\mathbf{p}
\end{aligned} \tag{6.103}$$

Substituting Eq. (6.101) into Eq. (6.103) and simplifying gives

$$\mathbf{P}_K = p_{\text{ideal}}\mathbf{I} - \int (\mathbf{p} - \mathbf{v}_0)(\mathbf{p} - \mathbf{v}_0)\left(\mathbf{B} : \frac{\partial \mathbf{v}_0}{\partial \mathbf{r}}\right) f_{(0)} d\mathbf{p} \tag{6.104}$$

where $p_{\text{ideal}} = nT$ is the (dimensionless) ideal gas pressure. Using the tensor form of **B** given by Eq. (6.92), the above equation can be rewritten as (Prob. 6.4).

$$\mathbf{P}_K = p_{\text{ideal}}\mathbf{I} - 2\mu\mathbf{S} \tag{6.105}$$

where **S** is the rate-of-strain tensor, also obtained previously for dense gases and liquids [cf. Eq. (6.45)]

$$\mathbf{S} = \frac{1}{2}\left[\left(\frac{\partial \mathbf{v}_0}{\partial \mathbf{r}}\right) + \left(\frac{\partial \mathbf{v}_0}{\partial \mathbf{r}}\right)^t - \frac{2}{3}\left(\frac{\partial}{\partial \mathbf{r}} \cdot \mathbf{v}_0\right)\mathbf{I}\right] \tag{6.106}$$

and the dimensionless viscosity coefficient μ is given in terms of the function B as

$$\mu = \left(\frac{1}{15}\right)\frac{1}{2T}\int B(w)|\mathbf{p} - \mathbf{v}_0|^4 f_{(0)} d\mathbf{p} \tag{6.107}$$

Equation (6.105) is recognized as a statement of Newton's law of viscosity, and the solution of the integral equation for the function $B(w)$, Eq. (6.94), for a specified intermolecular interaction potential, leads to specific values of the viscosity coefficient.[4,5]

We note that, due to the presence of intermolecular interactions, a potential contribution to the pressure tensor must also exist.† Enskog[6] was able to at least partly account for this contribution within the somewhat awkward framework of the Boltzmann transport equation methodology. We will not discuss Enskog's approach or more contemporary generalizations, and we simply refer the interested reader to "Kinetic Molecular Theory of Gases" (in the Further Reading section of Chap. 3) and the Further Reading section at the end of this chapter. In short, however, Enskog's modifications do not affect the basic functional form of the property flux expressions.

We now turn to the energy flux vector expression for gases. The dimensionless, kinetic contribution to the energy flux vector follows from

†From our scaling arguments (see the third footnote of Sec. 6.3), we have $P_{ko} = mn_0v_0^2 = mv_0^2/l_0^3$ and $P_{\phi_0} = mn_0^2v_0^2l_0^3 = mv_0^2/l_0^3$; thus, the potential contribution should be of similar order of magnitude as the kinetic contribution!

Eq. (5.66) as†

$$\mathbf{q}_k = \frac{1}{2}\int (\mathbf{p}-\mathbf{v}_0)^2\,(\mathbf{p}-\mathbf{v}_0)\,f^{(1)}(\mathbf{r},\mathbf{p},t)\,d\mathbf{p}$$
$$= \frac{1}{2}\int (\mathbf{p}-\mathbf{v}_0)^2\,(\mathbf{p}-\mathbf{v}_0)\,f_{(0)}(1+\phi)\,d\mathbf{p} \qquad (6.108)$$

Substituting Eqs. (6.101) and (6.102) into the above equation and simplifying gives (Prob. 6.5)

$$\mathbf{q}_k = -\lambda\frac{\partial T}{\partial \mathbf{r}} \qquad (6.109)$$

where the dimensionless coefficient of thermal conductivity λ is given by

$$\lambda = \frac{1}{3}(2T)^{1/2}\int A(w)\left(\frac{5}{2}-w^2\right)w^2 f_{(0)}\,d\mathbf{p} \qquad (6.110)$$

Equation (6.109) is known as Fourier's law of heat conduction. As noted above, there will be a potential contribution as well. Solution to the integral equation, Eq. (6.95), for the function $A(w)$ gives numerical values of the thermal conductivity from Eq. (6.110).[4,5]

Finally, we turn to the equation of entropy conservation and look at the specific expressions for the entropy flux and entropy generation terms. With entropy defined in the $s = 1$ space, the entropy flux from Eq. (5.86) in dimensionless terms is‡

$$\mathbf{s} = -\int (\mathbf{p}-\mathbf{v}_0)\,f_1 \ln(\hbar^3 f_1)\,d\mathbf{p} \qquad (6.111)$$

Substituting $f_1 = f_{(0)}(1+\phi)$, noting that

$$\ln[\hbar^3 f_{(0)}(1+\phi)] \cong \ln\left(\hbar^3 f_{(0)}\right) + \phi \qquad (6.112)$$

and finally using Eqs. (6.96) and (6.97), we obtain to linear order in ϕ

$$\mathbf{s} = -\int (\mathbf{p}-\mathbf{v}_0)\phi f_{(0)} \ln\left(\hbar^3 f_{(0)}\right)d\mathbf{p} \qquad (6.113)$$

Using Eq. (6.85) for $f_{(0)}$, we further have

$$\ln\left(\hbar^3 f_{(0)}\right) = \ln\left[\frac{n_1\hbar^3}{(2\pi T)^{3/2}}\right] - \frac{(\mathbf{p}-\mathbf{v}_0)^2}{2T} \qquad (6.114)$$

†Here, we let $q^* = q/q_0$, where $q_0 \equiv mn_0v_0^3$.

‡$\mathbf{s}^* = \mathbf{s}/\mathbf{s}_0$, where $s_0 \equiv kn_0v_0$. Also, note $\hbar^{*3}$ is a dimensionless Planck's constant cubed $[\hbar^{*3} = \hbar^3/(n_0 p_0^{-3})]$.

When substituted into Eq. (6.113), the first term of Eq. (6.114) vanishes by virtue of Eq. (6.97) leaving us with

$$\mathbf{s} = \frac{1}{2T}\int (\mathbf{p}-\mathbf{v}_0)(\mathbf{p}-\mathbf{v}_0)^2 f_{(0)}\phi\, d\mathbf{p}$$
$$= \mathbf{q}_k/T \tag{6.115}$$

by Eq. (6.108). Thus, we have derived the well-known thermodynamic argument that the entropy flux is equal to the energy flux divided by temperature.

For the entropy generation, we have from Eq. (5.88) in dimensionless terms[†]

$$n\bar{s}_g = -\int\int \ln\left(\hbar^3 f_1\right)\left[\frac{\partial\phi(\mathbf{r}_1,\mathbf{r}_2)}{\partial\mathbf{r}_1}\cdot\frac{\partial f_2}{\partial\mathbf{p}_1}\right] d\mathbf{r}_2 d\mathbf{p}_1 d\mathbf{p}_2 \tag{6.116}$$

In terms of the Boltzmann collision integral, we can write

$$\int \frac{\partial\phi}{\partial\mathbf{r}_1}\cdot\frac{\partial f_2}{\partial\mathbf{p}_1}\, d\mathbf{r}_2\, d\mathbf{p}_2 = \int\int\int v_{12}[f_1(1')f_1(2') - f_1(1)f_1(2)]b\, db\, d\varepsilon\, d\mathbf{p}_2 \tag{6.117}$$

Now, expanding as before [see Eqs. (6.83) and (6.84)], retaining terms to linear order in ϕ, and finally using Eq. (6.88), we get

$$n\bar{s}_g = \int\left[\ln\left(\hbar^3 f_{(0)}\right) + \phi\right] f_{(0)}\left\{\mathbf{b}_1 : \frac{\partial\mathbf{v}_0}{\partial\mathbf{r}_1} - \left(\frac{5}{2} - w_1^2\right)\right\}$$
$$\times\left[(\mathbf{p}_1 - \mathbf{v}_0)\cdot\frac{\partial\ln T}{\partial\mathbf{r}_1}\right] d\mathbf{p}_1 \tag{6.118}$$

Substituting our expression for ϕ from Eq. (6.91) and carrying out the integrations, we are left with (Prob. 6.6)

$$n\bar{s}_g = \int\left(\mathbf{B}_1 : \frac{\partial\mathbf{v}_0}{\partial\mathbf{r}_1}\right) f_{(0)}\left[2\left(\mathbf{w}_1\mathbf{w}_1 - \frac{1}{3}w_1^3\mathbf{I}\right) : \frac{\partial\mathbf{v}_0}{\partial\mathbf{r}_1}\right] d\mathbf{p}_1$$
$$- \int\left(\mathbf{A}_1\cdot\frac{\partial\ln T}{\partial\mathbf{r}_1}\right) f_{(0)}\left(\frac{5}{2} - w_1^2\right)\left[(\mathbf{p}_1 - \mathbf{v}_0)\cdot\frac{\partial\ln T}{\partial\mathbf{r}_1}\right] d\mathbf{p}_1 \tag{6.119}$$

Substituting Eqs. (6.92) and (6.93) for the functions $\mathbf{B}_1$ and $\mathbf{A}_1$, respectively, and finally using Eqs. (6.106), (6.107), (6.109), and (6.110) we obtain the desired result for entropy generation in the gas state as

$$n\bar{s}_g = \frac{1}{T}\left(\frac{\partial\mathbf{v}_0}{\partial\mathbf{r}} : 2\mu\mathbf{S}\right) - \frac{1}{T^2}\left(\mathbf{q}_k\cdot\frac{\partial T}{\partial\mathbf{r}}\right) \tag{6.120}$$

[†] $s_g^* = s_g/s_{g_0}$, where $s_{g_0} = n_0 k v_0 l_0^2$.

This result is well-known from phenomenological arguments,[7] and it is gratifying that we have obtained it from molecular theory. Note that entropy generation, via Eq. (6.120) is necessarily a positive quantity, which is truly a remarkable result that has critical consequences in setting limits on the exchange of heat and work in any given system. (See the footnote of Interlude 5.2 of Chap. 5.) Newton's law of viscosity, Fourier's law of heat conduction, and the positive entropy generation expression obtained here all indicate a specific direction of change for material processes, i.e., momentum flux and energy flux are always in a direction against velocity and temperature gradients, respectively, and entropy generation is always a positive quantity. All of these results constitute what is called *irreversible thermodynamics*. It remains to be seen whether or not the results given here can be carried over to the molecular theories of the liquid state given earlier in this chapter, or to highly nonequilibrium systems that cannot be represented by regular perturbations from local equilibrium states.

In the final section of this chapter we summarize our results, presenting the final closed forms of the transport equations. We note that extension of the molecular theory for gases can be extended to gas mixtures in a straightforward manner. (See the Further Reading section at the end of this chapter; also see Sec. 4.8.)

6.7 Chapter Summary and the Closed Forms of the Transport Equations

Using scaling analysis and perturbation methods, we have been able to derive approximate expressions for the momentum and energy flux in dilute gases and liquids. These methods physically involve formal expansions about *local* equilibrium states, and the particular asymptotic restrictions have been formally obtained. The flux expressions now involve the dependent transport variables of mass or number density, velocity, and temperature, and they can be utilized to obtain a closed set of transport equations, which can be solved simultaneously for any particular physical system. The problem at this point becomes a purely mathematical problem of solving a set of coupled nonlinear partial differential equations subject to the particular boundary and initial conditions of the problem at hand. (Still not a simple matter; see interlude 6.2.)

In this final section, and for ease of future applications, we return to dimensional variables and summarize the property conservation equations, or transport equations, using the results from Chap. 5 and Sec. 6.6. These sets of closed form equations are commonly used for both gas and liquid systems; albeit in liquids, caution must be exercised since Newton's and Fourier's law are used rather empirically, and violations, such as so-called non-newtonian behavior, are well-known. For alternative flux expressions, the general results in Chap. 5 can be used as a starting point to obtain particular forms.[7]

Mass conservation:

$$\frac{\partial n(\mathbf{r},t)}{\partial t} + \frac{\partial}{\partial \mathbf{r}} \cdot n(\mathbf{r},t)\mathbf{v}_0(\mathbf{r},t) = 0 \tag{6.121}$$

Momentum conservation:

$$mn\left[\frac{\partial \mathbf{v}_0}{\partial t} + \left(\mathbf{v}_0 \cdot \frac{\partial}{\partial \mathbf{r}}\right)\mathbf{v}_0\right] = -\frac{\partial}{\partial \mathbf{r}} \cdot p_{eq}\mathbf{I} + \frac{\partial}{\partial \mathbf{r}} \cdot (2\mu\mathbf{S}) + n\mathbf{F}_{\text{ext}} \tag{6.122}$$

Energy conservation:

$$n\left[\frac{\partial \bar{U}}{\partial t} + \mathbf{v}_0 \cdot \frac{\partial}{\partial \mathbf{r}}\bar{U}\right] = -\lambda\frac{\partial}{\partial \mathbf{r}} \cdot \frac{\partial T}{\partial \mathbf{r}} - p_{eq}\left(\frac{\partial}{\partial \mathbf{r}} \cdot \mathbf{v}_0\right) + \left(2\mu\mathbf{S} : \frac{\partial \mathbf{v}_0}{\partial \mathbf{r}}\right) \tag{6.123}$$

Entropy conservation:

$$\frac{n}{k}\left[\frac{\partial \bar{S}_{BG}}{\partial t} + \mathbf{v}_0 \cdot \frac{\partial \bar{S}_{BG}}{\partial \mathbf{r}}\right] = -\lambda\frac{\partial}{\partial \mathbf{r}} \cdot \frac{\partial \ln T}{\partial \mathbf{r}} + \frac{1}{T}\left[\frac{\partial \mathbf{v}_0}{\partial \mathbf{r}} : (2\mu\mathbf{S})\right] + \lambda\frac{1}{T^2}\left(\frac{\partial T}{\partial \mathbf{r}} \cdot \frac{\partial T}{\partial \mathbf{r}}\right) \tag{6.124}$$

Interlude 6.2 The fascinating world of nonequilibrium phenomena† The transport equations (mass, momentum, energy, and entropy) derived in this chapter mathematically represent a set of highly coupled, nonlinear partial differential equations, which, in practice, are difficult to solve. The nonlinearities of these equations give rise to a number of interesting behaviors, including spatial and temporal "chaos," turbulent fluid flows, quasi-periodicity, intermittency, and many other collective phenomena that are a direct consequence of nonlinearity.

The transport equations, however, may be linearized and solved as illustrated, for example, in Chap. 5 for the isothermal sound speed.[8] In these linearization methods, the density and/or temperature are expanded about a global equilibrium state. Such approaches fail when density and temperature changes are large, as in the case of two-phase systems (see Fig. 6.6).

So-called *order parameter* models are being currently used to understand a wide variety of systems that evolve outside of linearization regimes. These models can be motivated in several different ways, but we do not know yet of any systematic derivation that starts from first principles. Whereas linearized transport equations are well understood, their nonlinear extension

†Contributed by Jorge Vinals.

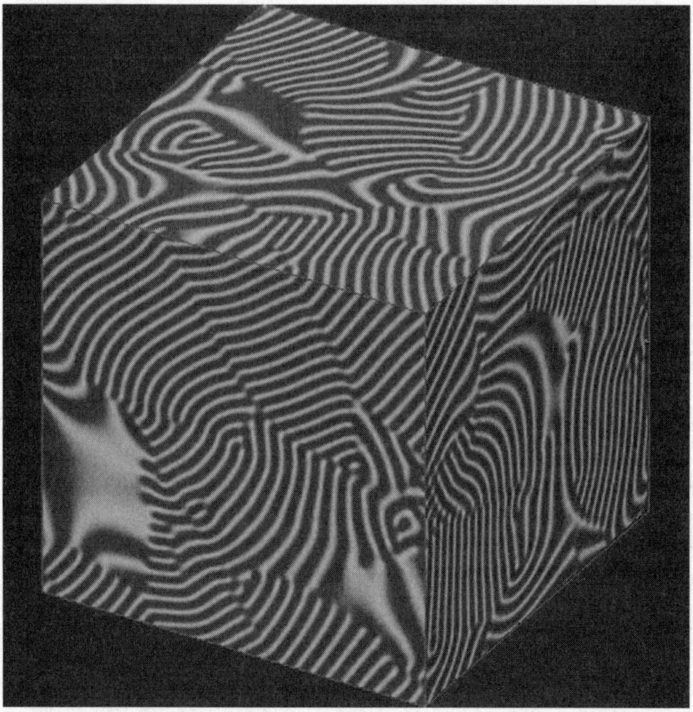

Figure 6.6 Instantaneous configuration (in grayscale) of the order parameter for microphase separation of a diblock copolymer AB forming a lamellar phase. This is a transient configuration from an initial, unstable uniform state, as the system evolves toward its final equilibrium configuration.

remains an open issue in many fields, including quantum optics, polymer physics, solidification, reaction-diffusion systems, or liquid crystals.[9]

The order parameter was arguably[†] first introduced by Landau at the equilibrium thermodynamic level to study phase behavior. Order parameter models can also be motivated through classical bifurcation theory, and several physical systems have been recently modeled in this way.[12] They include Rayleigh-Bénard convection,[13] Faraday waves,[14] or pattern formation in optical systems.[15] Close to a bifurcation point, these models asymptotically describe the system under study, but they are now routinely used, in a phenomenological fashion, to describe highly nonlinear phenomena.

[†]Historians debate this, as the metallurgists claim that the Cahn-Hilliard equation predated Landau. Each party then produces obscure conference reports, and I have never been able to make sense of this. It seems that the first one was used to describe superconductivity (see Ref. 9). Cahn introduced it for spinodal decomposition in Ref. 10. It is debatable whether the version for superconductivity ought to claim credit for its use outside the field.

Perhaps a sign of the growing importance of this approach is a recent review article in the journal *Reviews of Modern Physics* under the title "The world of the complex Ginzburg-Landau equation" (by I. Aranson and L. Kramer), just on solutions and properties of a particular type of order parameter model.

Problems

6.1 Using the chain rule for transformations, as in Eq. (6.7), derive the dimensionless RLE, Eq. (6.8), from Eq. (6.1).

6.2 Show that Eqs. (6.26) and (6.27) follow from substituting Eqs. (6.22) through (6.25) into Eq. (6.8).

6.3 Substitute Eqs. (6.86) and (6.87) into Eq. (6.84) to obtain Eq. (6.88). *Hint:* Refer to the development of Eq. (6.44) in the liquid state. The transport equations for equilibrium flows must be utilized.

6.4 In obtaining Newton's law of viscosity, Eq. (6.105), from Eq. (6.104), the use of the following integral theorem is made

$$\int f(c)\mathbf{cc}\left[\left(\mathbf{cc}-\frac{1}{3}c^2\mathbf{I}\right):\mathbf{d}\right]d\mathbf{c}$$

$$=\frac{2}{15}\left[\frac{1}{2}(\mathbf{d}+\mathbf{d}^t)-\frac{1}{3}Tr(\mathbf{d})\mathbf{I}\right]\int f(c)c^4d\mathbf{c}$$

where $\mathbf{d}$ is a tensor independent of $\mathbf{c}$. Prove this integral theorem.

6.5 In obtaining Fourier's law of heat conduction, Eq. (6.110), the use of the following integral theorem is made[†]

$$\int f(c)(\mathbf{a}\cdot\mathbf{c})\mathbf{c}\,d\mathbf{c}=\frac{1}{3}\mathbf{a}\int f(c)c^2d\mathbf{c}$$

where $\mathbf{a}$ is a vector independent of $\mathbf{c}$. Prove this integral theorem.

6.6 Show that Eq. (6.119) follows from Eq. (6.118).

References

1. A. Papoulis, *Probability, Random Variables and Stochastic Processes*, McGraw-Hill, New York, 1965.
2. M. Born and H. S. Green, A general kinetic theory of liquids III. Dynamical properties, *Proc. Roy. Soc. A*, **190**, 455, 1947; ibid, **188**, 10, 1946; ibid **189**, 103, 1947. [Max Born was awarded the 1954 Nobel Prize in Physics for his many significant contributions to quantum and statistical physics.]

[†]See Sec. 1.421 of Ref. 4.

3. N. Bogolubov, *Studies in Statistical Mechanics*, Vol. 1, J. de Boer and G. E. Uhlenbeck, (Eds.), North-Holland, Amsterdam, 1962.
4. S. Chapman and T. G. Cowling, *The Mathematical Theory of Non-Uniform Gases*, 3rd ed., Cambridge University Press, Cambridge, 1970.
5. J. O. Hirschfelder, C. F. Curtiss, and R. B. Bird, *Molecular Theory of Gases and Liquids*, Wiley, New York, 1964.
6. R. B. Bird, W. E. Stewart, and E. N. Lightfoot, *Transport Phenomena*, Wiley, New York 1960, 350.
7. J. P. Boon and S. Yip, *Molecular Hydrodynamics*, Chap. 5, McGraw-Hill, New York, 1980.
8. M. C. Cross and P. C. Hohenberg, Pattern formation outside of equilibrium, *Rev. Mod. Phys.*, **65**, 851 (1993).
9. V. L. Ginzburg and L. D. Landau, *Zh. Eksp. Teor. Fiz.*, **20**, 1064, 1950.
10. J. Cahn, On spinodal decomposition, *Acta Metall*, 9, 795, 1961.
11. J. P. Gollub and J. S. Langer, Pattern formation in nonequilibrium physics, *Rev. Mod. Phys.*, **71**, S383, 1999.
12. J. Swift and P. C. Hohenberg, Hydrodynamic fluctuations at the convective instability, *Phys. Rev. A*, **15**, 319, 1977.
13. W. Zhang and J. Vinals, Secondary instabilities and spatiotemporal chaos in parametric surface waves, *Phys. Rev. Lett.*, **74**, 690, 1995.
14. T. Ackeman and W. Lange, Optical pattern formation in alkali metal vapors: Mechanisms, phenomena and use, *Appl. Phys. B*, **72**, 1, 2000.

Further Reading

Scaling and Perturbation Methods

C. C. Lin and L. A., Segel, *Mathematics Applied to Deterministic Problems in the Natural Sciences*, MacMillian, New York, 1974.

J. Kervorkian and J. D., Cole, *Perturbation Methods in Applied Mathematics*, Springer-Verlag, New York, 1981.

M. Van Dyke, *Perturbation Methods in Fluid Mechanics*, The Parabolic Press, Stanford, CA, 1975.

A. Nayfeh, *Perturbation Methods*, Wiley, New York, 1973.

S. A. Rice and P. Gray, *The Statistical Mechanics of Simple Liquids*, Wiley, New York, 1965. [See this book for approximate methods in liquid state kinetic theory comprehensive text.]

E. A. Frieman, *J. Math. Phys.*, **4**, 410, 1963; C. H. Su, *J. Math. Phys.*, **5**, 1273, 1964. [See this book for more discussion and examples of multiple time-scale methods applied to reduced forms of the Liouville equation.]

You have brains in your head.
You have feet in your shoes.
You can steer yourself
any direction you choose.
You're on your own. And you know what you know.
And you are the person who'll decide where to go.

—THEODOR SEUSS GEISEL,
Oh, the Places You'll Go!

Index